Engineering Materials

This series provides topical information on innovative, structural and functional materials and composites with applications in optical, electrical, mechanical, civil, aeronautical, medical, bio- and nano-engineering. The individual volumes are complete, comprehensive monographs covering the structure, properties, manufacturing process and applications of these materials. This multidisciplinary series is devoted to professionals, students and all those interested in the latest developments in the Materials Science field, that look for a carefully selected collection of high quality review articles on their respective field of expertise.

Indexed at Compendex (2021) and Scopus (2022)

Yuli Panca Asmara

Concrete Reinforcement Degradation and Rehabilitation

Damages, Corrosion and Prevention

Springer

Yuli Panca Asmara
Department of Mechanical Engineering
Faculty of Engineering and Quantity
Surveying (FEQS)
INTI International University
Nilai, Negeri Sembilan, Malaysia

ISSN 1612-1317 ISSN 1868-1212 (electronic)
Engineering Materials
ISBN 978-981-99-5932-7 ISBN 978-981-99-5933-4 (eBook)
https://doi.org/10.1007/978-981-99-5933-4

This Springer imprint is published by the registered company Springer Nature Singapore Pte Ltd.
The registered company address is: 152 Beach Road, #21-01/04 Gateway East, Singapore 189721, Singapore

Paper in this product is recyclable.

Preface

Reinforced concrete is a construction material that combines concrete and steel reinforcement. Concrete is a strong and durable material, while steel provides high tensile strength. By combining these two materials, reinforced concrete is created, which has excellent structural properties. Concrete is also durable construction material that is extensively utilized in a wide range of applications. Its versatility allows it to be used in various construction projects and infrastructure developments. However, reinforced concrete can be susceptible to various types of damage, including corrosion of the steel reinforcement, carbonation, alkali-aggregate reaction, and surface deterioration.

Reinforced concrete can be susceptible to various forms of damages. Corrosion of the steel reinforcement is one of the most common forms of damage in reinforced concrete. When the passive film on the steel is compromised, it becomes susceptible to corrosion in the presence of water, oxygen, and chloride ions. This corrosion can lead to cracking, spalling, and loss of structural integrity. Carbonation is another form of concrete damage that occurs when carbon dioxide from the atmosphere reacts with the alkaline compounds in the concrete. Alkali-aggregate reaction (AAR) is a chemical reaction between the alkaline cement paste and certain reactive minerals in aggregates. Surface deterioration of concrete can occur due to various factors, including abrasion, erosion, freeze-thaw cycles, chemical attack, and weathering. These factors can cause cracking, scaling, and spalling of the concrete surface, affecting both the aesthetics and structural integrity of the concrete.

To detect and assess the damage in concrete, various inspection and testing methods can be employed. These include visual inspection, sounding, cover meter, half-cell potential measurement, chloride ion penetration test, carbonation test, and core sampling. These methods help in identifying the extent and severity of damage, which in turn informs the appropriate repair and maintenance strategies.

A thorough understanding of the causes and mechanisms of concrete damage is essential for implementing effective preventive measures and maintenance strategies. By adhering to proper construction practices, utilizing high-quality materials, employing protective measures, and conducting regular inspections, the durability and lifespan of reinforced concrete structures can be greatly enhanced. Additionally, additional preventive treatments, such as ensuring adequate concrete cover, incorporating corrosion-resistant steel, and applying protective coatings or implementing cathodic protection systems, can be implemented to further mitigate potential damages.

Nilai, Malaysia Yuli Panca Asmara

Acknowledgments This book would have been compiled with the assistance of many parties. We would like to express our gratitude to *Ap. Tutut Herawan, Prof. Wong Ling Shing, Dr. Chan Siew Chong, and Dr. Keng Hoo Chuah*, as well as all the staff at FEQS, Inti International University, Malaysia, for their support that made this book possible. We would also like to thank my family member:

Nunuk Ariyani,
Ferro Handaru Adidarma, and
Pascal Adiwidya Adiluhur.

Contents

About the Author

Assoc. Prof. Dr. Yuli Panca Asmara is a lecturer in Mechanical Engineering Department at INTI International University, Malaysia. He served as a postdoctoral fellow in the field of corrosion at University Technology Petronas (UTP). He was awarded a Ph.D. scholarship program from Malaysia National Oil Company (Petronas) to study corrosion in oil and gas environment. His master's degree in Corrosion Science was obtained from University of Manchester, UK. As an educational and research staff, he is conducting teaching and research activities of corrosion and materials science. More than 50 journals papers, inventions, and books references related with corrosion have been published. He is recognized as a senior professional engineer which allows him to conduct consultancy activates. Some consultancy projects had been carried out which were: a corrosion investigator in local companies, a trainer for corrosion problem in petroleum industries, and providing services for turnaround of Abu Dhabi Gas Liquefaction Company Limited (ADGAS).
e-mail: yuli.pancaasmara@newinti.edu.my

Chapter 1
Introduction to Reinforced Concrete

Abstract Reinforced concrete, with its remarkable strength and durability, has become a foundation in the construction industry. It is a composite material that combines the compressive strength of concrete with the tensile strength of steel reinforcement, making it capable of withstanding diverse loads and stresses. However, despite its inherent resilience, reinforced concrete is not impervious to damage. It is susceptible to a range of factors that can compromise its structural integrity and longevity. Environmental conditions, such as exposure to moisture, aggressive chemicals, and temperature variations, can trigger corrosive reactions and weaken the concrete and steel components. Additionally, excessive loads, whether from static or dynamic forces, can strain the material and lead to cracks, deformation, or even collapse. Effective measures need to be implemented to mitigate potential damages and maintain its structural integrity over time. In cases where damage has already occurred, appropriate treatments are crucial to restore and strengthen the reinforced concrete. The selection and application of suitable repair materials, such as epoxy resins or polymer-modified mortars, play a pivotal role in restoring the structural integrity and extending the service life of the concrete. In this context, this chapter raises discussions with the aims to explore the various aspects of reinforced concrete, a comprehensive understanding of the potential risks and measures of protection that it might face.

1.1 Reinforced Concrete

Reinforced concrete is a construction material that combines concrete and steel reinforcement. Concrete is a material with high hardness and compressive strength. Meanwhile, steel is a metal with high tensile strength and ductility. Combining these two materials produces a material with high tensile and compressive strength known as reinforced concrete. Concrete is made of gravel, sand, cement, and water. Cement functions as an adhesive that binds sand and stone together. Cement contains the main chemical elements of calcium, silica, and aluminum oxide. Water is needed for cement to turn into a solid. In concrete, steel becomes passive because concrete

Y. P. Asmara, *Concrete Reinforcement Degradation and Rehabilitation*,
Engineering Materials, https://doi.org/10.1007/978-981-99-5933-4_1

has a high pH. Furthermore, concrete can protect steel from interacting with the surrounding environment.

The strength of reinforcing steel is highly dependent on the quality of the concrete [1]. If the concrete contains active chemicals, the concrete becomes acidic, which can corrode the reinforcing steel. The quality of the concrete is determined by the mixed concrete ingredients and impurities. Unwanted elements in concrete, namely chlorine ions, sulfates, carbon dioxide (CO_2), and water can affect the quality of concrete. The element of chlorine contained in concrete comes from water vapor from the sea, which enters the concrete through its porous surface. Water is also introduced during the manufacturing of concrete using seawater as a mixer. Cement mixtures have the potential to damage concrete if sulfates, alkalis, or acids are present. Alkali and silica elements in concrete can be the cause of alkali-silica reaction [2–4]. This reaction causes the formation of sodium silicate gel ($Na_2SiO_3 \cdot nH_2O$). This hygroscopic gel increases in volume when it absorbs water and creates compression stress in the concrete, which reduces the chemical bonds in concrete and decreases its strength. Concrete can be damaged by external factors, such as excessive workload and failure during mixing, which can result in porosity, cracking, and collapse. Factors including corrosion in reinforcing steel, rapid heating during hydration, alkaline reaction (alkali-aggregate reactivity), CO_2 attack, acid attack, and chemical attack are other causes of concrete damage [5].

In order to produce high-quality concrete, modifications are made by changing the chemical composition of concrete and cement. Concrete is modified by adding fine particles (e.g., silica fume and fly ash) to reduce porosity and increase concrete strength. Silica fume is a very fine material in the form of particles less than 1 μm in diameter or 100 times smaller than the average cement particles. The addition of silica fume can also prevent the entry of chlorine ions into the concrete, hence reducing the risk of corrosion of reinforcing steel. The concretes are mixed with plasticizers to increase ductility. Plasticizers are added to concrete during manufacturing to reduce the water content in the concrete. Currently, concrete does not use cement as a binder, while resin is used as a substitute for cement [6, 7].

1.1.1 Damages of Reinforced Concrete

Low-quality concrete can damage the steel reinforcement in concrete. Steel is in a passive state because the concrete environment contains alkaline compounds. These compounds protect the steel by forming a thin film on the surface of the steel so that electrochemical reactions do not occur. However, the presence of external chemical elements that enter the concrete can damage the passive film and make the reinforcing steel in contact with the active environment. The corrosion rate of reinforcing steel in concrete structures determines the durability of the building structure. Corrosion of steel reinforcement is the most common factor causing damage to concrete structures. Reinforcing steel has serious corrosion problems when exposed to water, salt, and CO_2. In industrial areas where CO_2, hydrogen sulfide, sulfur, and carbon are present, the rate of deterioration of concrete will be faster. Reinforcement corrosion damages concrete and reduces its strength. Common types of corrosion in reinforced concrete are uniform corrosion, galvanic corrosion, and pitting. Thus, the need to know the causes of concrete damage becomes an important element in designing concrete structures [5].

Concrete damage affects the service life of concrete structural buildings. Concrete, simultaneously, is under load and undergoes chemical reactions with its environment. The longer the structures are exposed to the such conditions, the more damages will occur. Efforts to maintain the integrity of concrete must be made so that sudden damage does not occur. Maintaining and repairing concrete is a crucial work. The life of concrete can be extended in two ways, which are conducted before the concrete is cast or after the concrete is used [8–11]. Various considerations need to be evaluated to select the method for reinforced concrete curing. Other than the cost factor and level of complexity, efficiency also needs to be considered [12].

1.1.2 Reinforced Concrete Protection

Prevention of concrete damage and increasing the quality of concrete can be conducted by preventing corrosion in the reinforcing steel. Efforts to prevent corrosion in reinforcing steel include cathodic protection, anodic protection, coating, and replacing carbon steel with stainless steel. Cathodic protection is a method that changes the electrical properties of metals. The metal with a high oxidation potential (anodic) initially is changed to cathodic by means of a negative current flowing from an external source. Meanwhile, anodic protection is carried out by flowing current in the opposite direction [13–15]. Coating is a simple and easy method to apply for protection. The success of the coating system in protecting steel can be seen from its ability to prevent contact between steel and chemical elements that cause corrosion. Coatings are classified based on the type of coating material used. An example of corrosion prevention with the principle of a barrier layer is coating reinforcing steel with plastic/insulation. It prevents the steel to contact with concrete that has become acidic. Another example of the simplest coating method is painting. The basic ingredients of paint are made of binders (resin), forming mechanical properties and appearance (extenders and pigments), and solvents. The three main components of paint determine its properties and characteristics. Better coating properties are obtained by the use of metal (galvanizing), ceramics, and composites as materials for coating. Coating are also determined by coating techniques, for instance, coating with vapor deposition systems, cladding, thermal coatings, electroplating, and physical vapor deposition.

Preventing corrosion in concrete reinforcement can also be done by modifying the surface of steel reinforcement so that it always becomes passive, which can be done by adding chemicals to the concrete. Chemicals that are added to the concrete are called inhibitors, which are chemicals in small amounts that can reduce the rate of corrosion. Inhibition is one of the easiest methods to reduce the corrosion rate.

Many inhibitors have been produced in the market to prevent the corrosion of steel reinforcement. The selection of inhibitors is based on several factors, such as ease of operation, efficiency, and durability of the passive layer formed. Environmentally friendly inhibitors are also available to avoid environmental impacts [16].

The need to reduce risk of structural damages can also be met by producing concrete with a thicker cover. The thicker the concrete cover, the stronger the concrete structure in bearing the load. Reducing hydration time can also improve the quality of concrete, which can reduce the possibility of unwanted pollutants entering the concrete. It is known that the speed of drying determines the strength of the concrete. Hence, by using alternative media, the speed of drying can be adjusted. The use of CO_2 during drying can change the composition of the concrete, which can improve its mechanical properties.

1.1.3 Reinforced Concrete Treatment

The cost of treating corrosion is estimated at US$2.5 trillion, which is equivalent to 3.4% of global GDP (2013) (NACE). If maintenance is carried out, there will be savings between 15 and 35%. These costs do not include individual safety or environmental impacts. Several industries have realized that if corrosion is not properly anticipated, a bigger impact will occur, and it will be more expensive. In order to achieve a high level of safety, it is necessary to carry out maintenance and monitoring thoroughly and include corrosion as an integrated management system. Thus, corrosion experts must have various insights covering various aspects.

Corrosion engineers are responsible for the technical implementation of projects in all corrosion-prone environments. The corrosion specialist must take several actions including:

a. Ensure that all metal equipment runs according to plan and avoid work process stoppage due to corrosion.
b. Create a corrosion protection system and provide input if damage occurs due to corrosion.
c. Able to analyze the failure of a metal structure and be able to take preventive measures in the event of corrosion.
d. Estimate the calculation of maintenance costs.
e. Evaluate innovative designs, understand the concept of testing materials, and simulate corrosion processes.
f. Reduce the risk of corrosion.
g. Document, plan, implement, and continuously improve the organization's ability to manage corrosion threats to protect assets.

References

1. Asmara, Y.P., Nor, M.I., Anwar, S., Sugiman, S.: Remaining strength prediction of reinforced steel bar concrete structure in seawater environment. AIP Conf. Proc. **2489**(1), 030036 (2022)
2. Bentur, A., Berke, N., Diamond, S.: Steel Corrosion in Concrete: Fundamentals and Civil Engineering Practice. Taylor & Francis Group (2011)
3. Zerfu, K., Ekaputri, J.J.: Review on alkali-activated fly ash based geopolymer concrete. Mater. Sci. For. **841**, 162–169 (2016)
4. Zhenghao, Z., Guojiao, Y., Fang, C., Fumei, L., Qingyu, R., Haiping, Z.: Re-alkalization effect experiment and a new re-alkalization, control model of carbonated concrete. Adv. Mater. Sci. Eng. **6213832**, 11 (2022)
5. Portland Cement Association.: Types and Causes of Concrete Deterioration. PCA R&D Serial No. 261. (2002)
6. Burduhos Nergis, D.D., Abdullah, M.M.A.B., Vizureanu, P., Tahir, M.F.M.: Geopolymers and their uses: review. In: IOP Conference Series: Materials Science and Engineering vol. 374, p. 012019 (2018)
7. Phoo-ngernkham, T., Sata, V., Hanjitsuwan, S.: High calcium fly ash geopolymer mortar containing Portland cement for use as repair material. Constr. Build. Mater. **98**, 482–488 (2015)
8. Bennet, J., Schue, T.J.: Chloride removal implementation guide. In: Strategic Highway Research Program National Research Council ((1993)
9. Czarnecki, L.: Polymer-concrete composites for the repair of concrete structures. MATEC Web Conf. **199**, 01006 (2018)
10. Matsumoto, U., Ashida, M.: Study on re-alkalization with electrolytes containing lithium ion. Int. J. Mod. Phys. B **17**, 1446–1451 (2003)
11. Bastidas, D.M., Cobo, O., González, J.A.: Electrochemical rehabilitation methods for reinforced concrete structures: advantages and pitfalls. Corros. Eng. Sci. Technol. **43**(3), 248–255 (2008)
12. Baxi, C.K.: Anti-Corrosive Treatment for Concrete Surfaces. Austin Macauley Publishers. INTIUC-ebooks on–12–11 04:17:52 (2022)
13. Fontana, M.G.: Corrosion Engineering, 3rd edn. McGraw-Hill, New York (1986)
14. Revie Uhlig, R.W.: Corrosion and Corrosion Control: An Introduction to Corrosion Science and Engineering, 3rd edn. Wiley, New York (1985)
15. Callister, W.D., Rethwisch, D.G.: Materials Science and Engineering: An Introduction. Department of Metallurgical Engineering, The University of Utah
16. Ahmed, E.S.J., Ganesh, G.M.: A comprehensive overview on corrosion in RCC and its prevention using various green corrosion inhibitors. Buildings **12**(10)

Chapter 2
Concrete Structure

Abstract This chapter provides an exploration of fundamental aspects of concrete structures. It begins with an examination of concrete composition, encompassing the various materials that contribute to its formulation. The different types of concrete are then discussed, highlighting their unique properties and applications. The role of cement is explored, including an overview of its diverse types and the purpose they serve in concrete production. Concrete properties are a crucial focus, encompassing factors such as strength, durability, workability, and permeability. The discussion also encompasses concrete defects, identifying common issues that may arise during construction or over the lifespan of a structure. The interaction between concrete and steel reinforcement is explored, emphasizing the vital role of reinforcing elements in enhancing structural integrity. This leads to an exploration of the relationship between steel and concrete, delving into the bond formed through electrochemical processes and the mechanisms of capillarity and mechanical adhesion.

2.1 Composition of Concrete

Concrete is a solid structural material that is widely used as a building and infrastructure material in various conditions. Based on its function, the use of concrete is very wide and can be used for various purposes. Concrete is needed due to its availability, ease of use, ease of adaptation, and can have long life. Concrete is a very strong alloy and has high compressive stress. The strength of concrete can match that of natural stone. The concrete composition consists of sand, gravel, and cement. When the mixture undergoes hydration with water, cement binds sand and gravel to form mud, which can be molded into various shapes according to the mold shape (see Fig. 2.1). The maximum strength of concrete occurs when the concrete is about 28 days old, where the mixture dries to form a solid [1]. The hydration reaction is exothermic (releases heat); hence, special treatment is required to avoid damage due to compaction. The weakness of concrete lies in high density, low tensile strength, ease of cracking, and difficulty to recycle.

In order to improve the weakness of concrete, additional ingredients are often added to the concrete in the form of pozzolans and superplasticizers to improve

Y. P. Asmara, *Concrete Reinforcement Degradation and Rehabilitation*,
Engineering Materials, https://doi.org/10.1007/978-981-99-5933-4_2

Fig. 2.1 Concrete composition of sand, cement, gravel, and water. Additives and fly ash modify the desired physical and chemical properties of concrete

the physical and mechanical properties, as well as drying speed. Superplasticizer is a type of additive that can reduce the need for water in producing concrete. By reducing the water content, the possibility of excessive hydration can be avoided. The added sand and gravel are used to increase the strength of the concrete. Other modifiable properties of concrete are ductility, hardness, and resistance to corrosion attack. Concrete with special properties can be produced by adding chemical substances to the concrete mix. Additional substances such as silica fume and fly ash are useful for shrinking concrete pores. In addition, the properties of concrete are also determined by the type of cement, type of gravel, water, additives, and the composition of each alloy. Various types of aggregates can be used to improve the properties of concrete, namely sand, gravel, crushed stone, and iron furnace slag. The type of aggregate determines workability, strength, and service life. Concrete that uses pumice aggregate will have the property of easily absorbing water and is lightweight, hence suitable for use in building partitions (see Fig. 2.2).

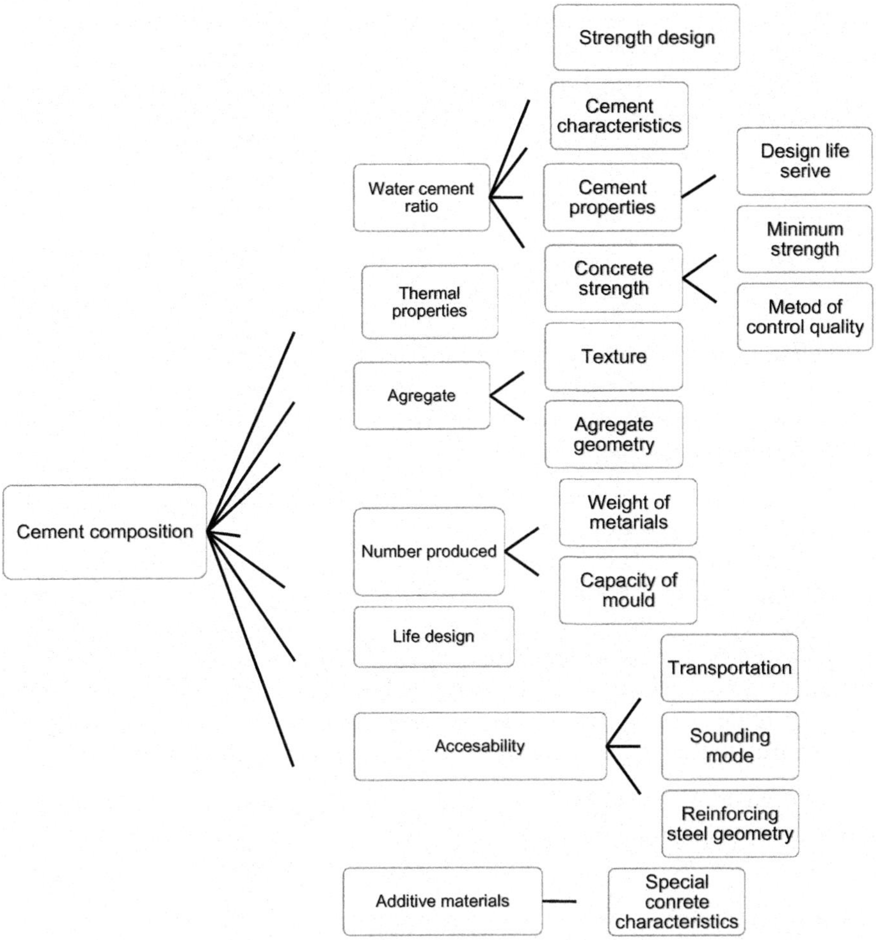

Fig. 2.2 Factors to be considered for making reinforced concrete [2]

2.2 Types of Concrete

The strength of a concrete is determined by the combination of its constituent elements. The main components of concrete are water, coarse aggregate, fine aggregate (sand), and air. In order to make concrete with the desired strength, a concrete design formula (mix design) is usually used, where the number and type of mixed elements are determined based on mathematical calculations. Basically, the concrete mix follows the guidelines (ACI Committee, 2009). Concrete can be designed using the tables and calculations provided in the standard. Table 2.1 is an example of designing concrete strength by considering the component ratios of each mixture.

Table 2.1 Mix design based on IS456:2000 to produce concrete with a certain strength by calculating the ratio of cement: sand, aggregate [3]

Type of concrete	Concrete grade	Mix ratio	Compressive strength after 28 days (N/mm^2)
Ordinary concrete	M5	1:5:10	5
	M7.5	1:4:8	7.5
	M10	1:3:6	10
	M15	1:2:4	15
	M20	1:1.5:3	20

The use of concrete in the field of infrastructure has become a necessity in all industries. Due to efficiency demands, concrete can be produced with different variations. Various types of concrete have been developed. Classification of concrete is based on various parameters, namely:

a. During manufacture: concrete with special treatment to produce certain properties and without special treatment.
b. Concrete weight: heavy concrete and light concrete. Lightweight concrete (specific gravity <1,900 kg/m^3) is produced from lightweight and porous aggregates.
c. Binder types: cement, polymer, and asphalt binders.
d. Applications: concrete used in water highways, prestressed concrete, precast concrete, decorative concrete, thermal insulation concrete (lightweight concrete), waterproof concrete, and soundproof concrete (lightweight concrete).

According to ACI 318 [4], the following Table 2.2 presents the requirements for the type of concrete used for various conditions.

Table 2.2 Requirements for the type of concrete used for various conditions [4]

Working conditions	Maximum w/cm (1)	Minimum compression strength (2)
Low permeability concrete, concrete in aqueous environments	0.50	4,000
Concrete in very cold conditions where it freezes and thaws	0.45	4,500
Concrete resistant to chloride attack, chemicals, salt, brackish water, and seawater	0.40	5,000

(1) The ratio of the weight of water to cement
(2) Normal and lightweight aggregate concrete

2.3 Cement

In a concrete, cement acts as a binder. Cement binds sand and gravel (aggregate) together. The main ingredients of cement are calcium and silicon, which in nature react with oxygen to form calcium silicate. Cement can be divided into two groups: cement that requires water to harden (hydraulics) and cement that does not need water but hardens in air and with carbon dioxide (CO_2) (non-hydraulics). The general chemical compositions of cement alloys are C_3S:alite ($3CaO{\cdot}SiO_2$), C_2S:belite ($2CaO{\cdot}SiO_2$), C_3A:tricalcium aluminate ($3CaO{\cdot}Al_2O_3$), and C4AF:brownmillerite ($4CaO{\cdot}Al_2O_3{\cdot}Fe_2O_3$) as described in Table 2.3 as follow.

Cement is made by mixing various types of main raw materials, namely metals and minerals, such as aluminum, iron, calcium, and silicon. Raw materials in cement may include lime, clay, iron ore, limestone, silica sand, and slag [7]. The material is heated at high temperatures to form a solid material called clinker. The clinker is then ground into powder. In cement making, tests are carried out to ensure that the cement produced meets industry standards. Cement quality tests that can be carried out include air content, compressive strength, strength during drying, temperature during hydration, particle size, and mixing time.

a. **Cement type**

In the construction industry, cement is differentiated based on its chemical composition, as well as its function and usage. There are 13 types of cement in the market, namely [8]:

Table 2.3 The chemical composition and physical properties of cement [5, 6]

Properties		Portland cement	Siliceous[b] fly ash	Calcareous[c] fly ash	Slag cement	Silica fume
Proportion by mass (%)	SiO_2	21.9	52	35	35	85–97
	Al_2O_3	6.9	23	18	12	—
	Fe_2O_3	3	11	6	1	—
	CaO	63	5	21	40	< 1
	MgO	2.5	—	—	—	—
	SO_3	1.7	—	—	—	—
Specific surface (m^2/kg)[d]		370	420	420	400	15,000–30,000
Specific gravity		3.15	2.38	2.65	2.94	2.22
General purpose		Primary binders	Cement replacement	Cement replacement	Cement replacement	Property enhancers

a Values shown are approximate; those of a specific material may vary
b ASTM C618 Class F
c ASTM C618 Class C
d Specific surface measurements for silica fume by nitrogen adsorption method, others by water permeability method

a. **Ordinary Portland cement (OPC)**
Ordinary Portland cement is a type of cement that is produced and used in the construction industry. This cement is widely used for concrete, mortar (adhesive), and building coatings. Portland cement can be classified into five types ranging from type 1 to type 5 based on its strength and use, which are: type 1 (OPC), type 2 (moderate sulfate resistance), type 3 (high early strength), type 4 (low heat of hydration), and type 5 (sulfate-resistant cement).
b. **Portland pozzolana cement (PPC)**
In order to make PPC, pozzolanic clinker is added to OPC. Portland pozzolana cement is resistant to chemical attack; thus, it is widely used in construction, including marine structures, sewage works, bridges, wharves, and dams.
c. **Fast-setting cement**
This type of cement hardens quickly and has high strength in the early stages of the setting process. Fast-setting cement has high lime content and fine aggregate.
d. **Sulfate-resistant cement**
Sulfate-resistant cement is made to reduce the risk of sulfate attack on concrete. The cement is used in locations containing high sulfate content.
e. **High alumina cement**
High alumina cement is a type of fast-setting cement that is made by adding bauxite and lime together and grinding them with clinker. This cement has high compressive strength and is more flexible in its use. High alumina cement is suitable for use in environments subject to high or low temperatures.
f. **White cement**
White cement is a type of OPC that is white in color and used for decorative and architectural purposes.
g. **Air-containing cement**
This cement has air bubbles to provide space if the concrete expands to prevent cracks and damage to the concrete.
h. **Expansive cement**
Expansive cement can grow slightly over time without shrinking during setting. This is beneficial for grouting anchor bolts or concrete drains.
i. **Hydrographic cement**
Hydrographic cement is made by mixing water-repellent chemicals. This type of cement is impermeable to water attack. It is suitable for dams, water tanks, overflow canals, and water-retaining structures.

2.4 Types of Cement Additives

Additives are added to cement to obtain additional properties according to the desired goals, such as ease of production, flowability, increased strength, and durability. Additives added to cement increase the efficiency of the mixing process, reduce water content, reduce pores, accelerate hardening, and increase strength. Cement additives are made of various chemical elements that can change the natural properties of

Table 2.4 Effect of the type of additives on the properties of concrete [9]

Additives	Composition	Functions	Content by weight of cement (%)
Bentonite	Montmorillonite (main mineral)	Reduces water vapor permeability	6
Calcium chloride	$CaCl_2$	Compaction catalyst	355
Liquid glass	$Na_2O\cdot(SiO_2)n$	Reducing the entry of water vapor	4
Melflux	Polyether carboxylate spray drying products	Plasticizers	0.75
S3	Polynaphthalene methylene sulfonate sodium salts	Strength booster	0.6
D5	Sodium sulfate, andesine, quartz, muscovite, kaolinite	Compaction catalyst, strength enhancer, frost resistant, moisture permeability reduction, plasticizer	2.25
Nordplast	Solution of complex salts	Solidification catalyst, frost resistant, plasticizer	0.6

cement with the aim of changing its characteristics. The success of adding additives to change the properties of cement depends on the dosage of the additive, type, method of use, design and operation, and cement characteristics.

There are many types of additives available in the market, which are usually made from a combination of organic and inorganic compounds.

- Triethanolamine- or alkanolamine-based additives facilitate mixing and increase cement hydration. These additives can catalyze C_3S hydration to produce high concrete strength.
- Na/Ca chloride and Na acetate/format are useful for increasing cement strength.
- Polymer-based additives can contribute to the fluidity of the mixture.

Table 2.4 as follow presents the effect of the type of additives on the properties of concrete.

2.5 Concrete Properties

a. Strength

   Concrete strength can be classified into several types as follows:

   a. Compressive strength.
   b. Tensile strength.
   c. Flexibility (flexural strength).
   d. Shear strength.

Compressive strength is the ability of concrete to withstand compressive loads perpendicular to the area before the concrete breaks. The unit of compressive strength is the force divided by the area of the concrete. Normal concrete strength varies between 25 and 40 MPa. Above 50 MPa, the concrete is classified as high-strength concrete (HSC). Meanwhile, tensile strength is the strength of concrete when subjected to a tensile load. Shear strength is the ability of concrete to withstand loads parallel to the fracture plane before the concrete fails. The unit for shear strength is the force divided by the area. If the force acting on the concrete exceeds the shear strength of the concrete, concrete cracks will occur. Shear strength is difficult to determine experimentally compared to other mechanical strength properties due to the difficulty of isolating shear [10]. Usually, the shear strength determined from various experimental studies shows a variation between 20 and 85% compared to compressive strength.

b. *Ease of Use (Workability)*
   Concrete must have properties that are easily moved with properties that do not change. Producing stable strength requires a consistent effort. The quality of the concrete must be obtained when the concrete is transported, placed, and finished without damage. A concrete that meets these conditions is said to have good workability properties, which can be influenced by water content; mix composition; aggregate size, shape, and texture; use of admixtures and additives; hydration time; and temperature.
c. *Concrete Durability*
   The ability of concrete to remain undamaged for a long time is called durability. The compressive strength of concrete is affected by its permeability to water and other potentially damaging materials. Good quality cement has a fairly low water/cement ratio.
d. *Chemical Resistance*
   Concrete is highly resistant to physicochemical attacks of environmental origin (e.g., frost and atmospheric pollution rain). Concrete is very suitable for structures that are exposed to extreme conditions (i.e., temperature, pressure, and corrosive).
e. *Porosity and Specific Gravity*
   The denser the concrete, the better it will perform in resisting chemical and water attack. Concrete density can be increased by optimizing dimensions and aggregate density, as well as reducing the moisture content. An important property of concrete is its specific gravity, which is related to the ease of being moved.
f. *Creeps*
   Creep is the growth of cracks in concrete or the shifting of cracks. Concrete that holds loads for a long time tends to creep. The amount of reinforcement and load that occurs in the concrete structure contributes to the size and number of cracks.
g. *Expansion*
   Concrete has a very low coefficient of thermal expansion. However, if temperature and pressure combine in concrete, the concrete will experience cracks in the weak parts of the structure. Concrete is unable to withstand repeated forces or cycles

of expansion and contraction. The coefficient of thermal expansion of Portland cement concrete is 8×10^{-6}–12×10^{-6} strains/°C [11].

h. *Elasticity*
Concrete cannot have perfect elasticity when subjected to loading. Elasticity is the ratio between load and deformation. The elasticity of concrete depends on the composition and the finish of the concrete.

2.6 Reinforced Concrete (Reinforcing Steel Bar–Rebar)

Although concrete is very strong, it is only strong in resisting compressive loads. Concrete has low strength in receiving tensile loads. Concrete is combined with steel, which is called reinforced concrete, to increase its ability to accept tensile loads. Reinforced concrete is concrete in which iron reinforcement is embedded to provide additional tensile strength. The alkaline chemical environment (KOH, NaOH, and CaOH) in the cement forms a passive film on the surface of the steel and makes it resistant to corrosion attack. This compound can protect steel surfaces against corrosion attack when the cement paste is exposed to air and water is polluted by CO_2 from the atmosphere, portlandite (*calcium hydroxide*), and hydrated calcium silicate. The attack by these chemical elements causes the cement paste to experience carbonation, and the pH will decrease to 8.5. The pH in the concrete changes to calcite (calcium carbonate) and the steel is no longer passive. The corrosion rate of steel for passive concrete is about 0.1 μm per year. If the steel is no longer passive, its corrosion rate is 1,000 times higher [12].

The corrosion rate of reinforcing steel in concrete structures determines the durability of the building structure. Corrosion of steel reinforcement is the most common factor causing damage to concrete structures. Reinforcing steel has serious corrosion problems when exposed to water, salt, and CO_2. In industrial areas where CO_2, hydrogen sulfide, sulfur, and carbon are present, the rate of deterioration of concrete will be faster. Reinforcement corrosion damages concrete and reduces its strength. Typical forms of corrosion in reinforced concrete are uniform corrosion, galvanic corrosion, and pitting. Thus, the need to know the causes of concrete damage is important in designing concrete structures. Factors that can cause concrete damage include:

- Steel corrosion.
- Rapid heating on hydration.
- Alkaline reaction (alkali-aggregate reactivity).
- CO_2 attack.
- Sulfate attack.
- Acid and chemical attack.

2.7 Concrete Defects

a. ***Concrete Porosity***
Concrete is a composite material consisting of coarse and fine aggregates in a cement mixture that undergoes hardening. Concrete is water resistant and resistant to chemical attack. However, if the concrete has pores, chemical elements from outside can enter the concrete. The properties of the cement paste, as well as the pore structure and composition of the solution, are the dominant factors in determining the quality of the concrete. The cement paste structure occurs as a result of the chemical reaction between cement and water. When the concrete is molded and dried, the volume of hydrated cement will produce a larger product, resulting in the formation of void spaces in the concrete. The number of voids formed is determined by the ratio of the amount of water to cement (w/c) and the time required for drying. When the w/c is low, the porosity that occurs will be smaller.

b. ***Shrinkage Cracks***
Shrinkage occurs due to concrete *volumetric* changes (shrinkage) during drying. When shrinkage occurs, concrete can experience cracks due to thermal effects and expansion. Therefore, during drying, it is necessary to hold the concrete so that the concrete-steel bonding capacity does not decline. Usually, a retainer is placed both inside the concrete and outside the form. Cracks caused by shrinkage are affected by the amount of shrinking volume, the retaining structure, and the number and spacing of reinforcement. Cracks in concrete can occur during drying, as well as when the concrete is in a set state.

c. ***Chemical Changes***
High environmental temperatures cause concrete to lose its hydration products due to water evaporation. Therefore, the resistance of concrete to water will decrease. The amount of unhydrated cement increases with the decreasing amount of water. As a result, the strength of the concrete will decrease [13]. Evaporation also results in the conversion of calcium hydroxide ($Ca(OH)_2$) in the concrete into lime and water. When the temperature decreases, the lime will react with water and expand, causing a decrease in strength [14].

d. ***Physical Changes***
At high temperatures, small cracks form and propagate in the concrete as the temperature increases. Cracks mainly occur at the junction of different components due to differences in the coefficient of thermal expansion of the components in the cement. Likewise, when water evaporates from the concrete, the loss of water inhibits the expansion of the cement. In addition, when the temperature reaches 573 °C, the silica aggregate changes from the α-phase, hexagonal crystal system to the β-phase, body-centered cubic structure, causing expansion of the concrete and a decrease in strength [15].

e. ***Spalling***
Peeling at high temperatures occurs due to vapor pressure and thermal stress [16]. When the concrete surface is exposed to sufficiently high temperatures, water close to the surface begins to move out of the concrete and into the atmosphere.

However, the high-temperature gradient between the surface and the inside of the concrete causes vapor to enter the concrete, which can cause condensation if the temperature drops. If the steam condensation rate is much faster than the steam exit rate from the concrete, flaking will occur. When thermal expansion occurs at the surface of the concrete, followed by stress opposite to the tensile stress within the concrete, spalling will occur. Spalling occurs when the compressive stress exceeds the tensile stress [17].

f. ***Temperature Changes***
Reinforced concrete consists of concrete and steel, each of which has a different coefficient of thermal expansion. Due to differences in thermal expansion and contraction, reinforced concrete is prone to excessive heat, reducing the bonding capacity of concrete to steel. Ideally, when the cement paste in the concrete hardens, any stress that the concrete receives can be transmitted to the steel. Usually, the steel bar is rough or corrugated to further increase the bond between the concrete and the steel.

2.8 Concrete-Steel Interaction

Concrete is an alkaline material that is formed from the hydration between cement and fillers in the form of stones, gravel, and sand. Several components (Ca^{2+}, OH^-, Na^+, K^+, and SO^{2-}) react with water on the concrete-steel surface, which is called the interfacial transition zone (ITZ). The hydration reaction affects the reinforcing steel embedded in the concrete, increasing the strength of the concrete. When the concrete compacts, this area becomes a gathering place for the alkaline layer of $Ca(OH)_2$, which determines the strength of the steel–concrete bond. At the ITZ, there are three types of layers due to the simultaneous occurrence of sedimentation, segregation, hydration, and mechanical vibration. The first layer is 10–100 μm from the steel reinforcement consisting of fine cement particles and water-cement mixture. The second layer with a thickness of 5 mm contains mortar, large sand grains, cement, and water. The last layer is concrete filled with gravel grains with a size of about 30 mm [18]. The adhesion between steel and concrete occurs due to the seepage of cement into the layer between sand and gravel, where the absorption of the water layer forms between the steel-cement and electrochemical bonds with $Ca(OH)_2$. The adhesive strength in the ITZ determines the strength of reinforced concrete.

The factors that determine the strength of the steel–concrete bond are as follows:

a. ***Steel–Concrete Electrochemical Bond***
Concrete is a material that has an active chemical structure. The interaction of concrete and metal reinforcement produces electrochemical and electrical bonds. This bond occurs due to intermolecular electromagnetic interactions that are formed during concrete compaction. Thus, a way to increase the steel–concrete adhesion (See Table 2.5) force is to improve the electrochemical properties along the location of the steel surface in the concrete. The steel–concrete bonding forces involved are intermolecular van der Waals forces (40 kJ·mol^{-1}), hydrogen

bonding forces (50 kJ·mol^{-1}), covalent bonds (60–700 kJ·mol^{-1}), metallic bonds (110–260 kJ·mol^{-1}), and ionic bonds (600–1,000 kJ·mol^{-1}) [18]. Therefore, the concrete mix, adsorption, adhesion, and carbonation kinetics during curing are the key factors to obtain a strong electrochemical bond. Cement contains large amounts of OH^- hydroxyl radicals. The electronegativity of iron in steel and oxygen in OH^- radicals allows hydrogen bonding between iron Fe atoms and OH radicals to form $Fe(OH)_2$ molecules, followed by the formation of an oxyhydroxide film [19].

b. ***Capillarity***
Capillarity is the capillary force that is caused by the adhesion and cohesion forces that determine the surface tension of the liquid. Surface tension in the steel acts along the steel–concrete surface interaction. Capillary forces are formed when the liquid film is along the ITZ. To obtain high adhesion strength on steel–concrete, the hydration process must take place quickly. Capillarity is also affected by contact angles, adsorption forces, covalent bonds, and hydrogen dipole bonds between concrete and metal.

Table 2.5 Steel–concrete adhesion to various steel and concrete surface conditions [16]

Reinforced bar		Concrete			Bond strength	
Material	Surface finish	Ultimate compressive stress (MPa)	Hardening conditions	Aging time		Fracture stress (MPa)
Bare steel	Smooth	39–67	20 °C/ 100% RH	1 day	Pull out	5–9
Bare steel	Smooth	88–100	20 °C/ 100% RH	28 days	Pull out	8–10
Bare steel	Rough	39–67	20 °C/ 100% RH	1 day	Pull out	13–18
Bare steel	Rough	88–100	20 °C/ 100% RH	28 days	Pull out	23–27
Bare steel	Smooth	15.6	–	3 days	Pull out	6–8
Bare steel	Smooth	31.3	–	28 days	Pull out	8–14
Galvanized steel	Smooth	15.6	–	3 days	Pull out	3–6
Galvanized steel	Smooth	31.3	–	28 days	Pull out	4–7
Bare steel	Smooth	14.6	–	3 days	Pull out	5–6
Bare steel	Smooth	29.2	–	28 days	Pull out	8–10
Galvanized steel	Smooth	14.6	–	3 days	Pull out	3–6
Galvanized steel	Smooth	29.2	–	28 days	Pull out	9–13

c. ***Mechanical Adhesion***
   The steel–concrete mechanical strength is formed when the concrete compacts, where the steel is held in place by the surface roughness and geometry of the steel (Table 2.5). The adhesive strength depends on the amount of area of steel adhering to the concrete and the surface texture of the steel. Hence, to increase the mechanical bond strength, the steel is designed to have a texture that causes surface defects to be filled with concrete. The flowability of concrete is required so that the concrete can fill all the spaces in the steel–concrete, which is usually done by adding superplasticizers. Surface roughness of steel, use of finned rods, water pressure, wall pressure, and mechanical vibration when molding concrete are the main factors that increase mechanical bond strength [20].

2.9 High-Quality Concrete

High-performance concrete (HPC) exceeds the properties and constructability of normal concrete. High-strength concrete has a minimum design strength of 70 MPa (10,000 psi). The specific materials used to produce this concrete must meet performance requirements. Special mixing, placing, and curing techniques are required to produce HPC. Performance tests are required to demonstrate compliance with field requirements [21, 22]. High-performance concrete is primarily used in tunnels, bridges, and high-rise buildings due to its high strength, durability, and modulus of elasticity. The characteristics of HPC are developed for specific applications and environments, and some of the required properties include [23]:

- High elastic modulus.
- High abrasion resistance.
- High durability and long life in harsh environments.
- Low permeability and diffusion.
- Resistance to chemical attack.
- Volume stability.
- Light.
- Compaction without segregation.
- Inhibits the growth of bacteria and fungi.

High-quality concrete has a w/c ratio between 0.20 and 0.45. Plasticizers are used to make concrete flexible and not easily cracked. Bickley and Fung [24], studied that the compressive strength of 40 MPa of HPC could be used to build environmentally friendly bridges. High-performance concrete is often prepared with the aim of resisting chloride attack. Table 2.6 shows the materials required to produce HSC.

High-strength concrete is made with the following requirements:

- The cement used must be of high quality.
- High cement content (400–600 kg/m^3).
- Low w/c ratio (0.20–0.45 by weight ratio).
- Higher concrete temperature.

Table 2.6 Materials used in HPC [25]

Material	Primary contribution/desired property
Portland cement	Cementing material/durability
Blended cement	Cementing material/durability/high strength
Fly ash	Cementing material/durability/high strength
Slag	Cementing material/durability/high strength
Silica fume	Cementing material/durability/high strength
Calcined clay	Cementing material/durability/high strength
Metakaolin	Cementing material/durability/high strength
Calcined shale	Cementing material/durability/high strength
Superplasticizers	Flowability
High-range water reducers	Reduce w/c ratio
Hydration control admixtures	Control settings
Retarders	Control settings
Accelerators	Accelerate setting
Corrosion inhibitors	Control steel corrosion
Water reducers	Reduce cement and water content
Shrinkage reducers	Reduced shrinkage
Alkali-silica reactivity inhibitors	Control alkali-silica reactivity
Polymer/latex modifiers	Durability
Optimally graded aggregates	Improve workability and reduce paste demand

- Higher curing temperature.
- Chemical mixing.
- Drying with steam or autoclave.
- Insulation to withstand the heat of hydration.
- Using fast-hardening cement.

a. ***High-Quality Concrete Material***

a. ***Cement***

The selection of cement for HSC should not only be based on the mortar-cube test but should also include concrete strength comparison tests at 28, 56, and 91 days of age. The cement must produce a mortar-cube strength of at least 7 days of 30 MPa (4,350 psi). The sand content is made as low as possible [25].

b. ***Supplementary Material***

Fly ash, silica fume, or slag can be used to make HSC. Additives are added at a rate of 5%–20% or higher per mass of cement. Some specifications only permit the use of up to 10% silica fume, unless there is evidence to suggest that concrete produced in larger doses will have satisfactory strength, durability, and volume stability. The cement-to-water ratio must be adjusted so that the same performance is the basis for comparisons between trial mixtures.

c. ***Aggregate***
In HSC, attention must be paid to aggregate size, shape, surface texture, mineralogy, and cleanliness. For each aggregate source and concrete strength level, there is an optimum aggregate size that will produce the greatest compressive strength per cement unit. Many studies have found that the aggregate nominal maximum size of 9.5–12.5 mm (3/8–1/2 inch) provides optimal strength. Tests have shown that crushed stone aggregate produces a higher compressive strength in concrete than gravel aggregate. This is because the aggregate-to-paste bond is better when using coarse, angular, and crushed materials [25].

d. ***Chemical Material***
The use of chemicals such as water reducers, retarders, or superplasticizers is required to produce HSC. These materials make the use of cement more efficient. The use of air-retaining chemicals is neither necessary nor desirable in HSC that is protected from weathering, such as interior columns and shear walls of high-rise buildings. However, for bridges, concrete piles, piers, or parking structures, which require durability in a freeze–thaw environment, air-breaking chemicals are necessary.

2.10 Properties of High-Quality Concrete

a. ***High-Durability Concrete***
The durability of concrete is because concrete has a low diffusion coefficient (4.8×10^{-13}). This value is 10–30 times lower than conventional concrete. Besides, the electrical resistance properties (electrical resistivity) must also be possessed by high-quality concrete. High-strength concrete has a resistance of 470–530 Ω-m, which is much higher than conventional concrete (50 Ω-m). High concrete resistivity results in a low corrosion rate or a 10% reduction in the corrosion rate of conventional concrete [26].

b. ***Abrasion Resistance***
Abrasion resistance is related to the strength of concrete. High-strength concrete is made using very high silica fume content. Silica fume is useful for concrete pavements, which are often used for highway pavements. The best results are obtained by mixing cement slag, steel fiber, and silica fume. Mortar strength ranges from 75 to 100 MPa. Abrasion-resistant highway mixes typically contain between 320 and 450 kg/m^3 of cement, plus silica fume or fly ash [25].

c. ***Explosion Proof***
High-strength concrete can be designed to have excellent explosion-resistant properties. This concrete often has a compressive strength exceeding 120 MPa (14,500 psi) and contains steel fibers. Explosion-resistant concrete is often used in bank vaults and military applications.

d. ***Permeability***

The durability and service life of concrete exposed to weather are determined by the permeability of the concrete. High-strength concrete has very low permeability to air, water, and chloride ions. The porous structure of HPC makes it highly impermeable to chemical attack. This concrete can be modified with latex and is able to achieve low permeability levels.

e. ***Diffusion***

Aggressive ions, such as chloride, in contact with the concrete surface, will diffuse through the concrete until an equilibrium state in ion concentration is reached. If the concentration of ions at the surface is high, diffusion can lead to concentrations that promote corrosion at the reinforcement level. The lower the w/c ratio, the lower the diffusion coefficient of the concrete. An additional ingredient in the form of silica fume is added to reduce the rate of diffusion.

f. ***Carbonation***

High-strength concrete has excellent resistance to carbonation due to its low permeability. High-strength concrete that is not cracked is immune to carbonation to depths, which will cause corrosion.

g. ***Chemical Attack***

High-strength concrete is resistant to chemical attack. Resistance to various sulfates is achieved using dense, strong concrete with very low permeability and a low w/c ratio.

h. ***Alkali-Silica Reactivity***

The reactivity between silica aggregates and alkaline hydroxides can affect the long-term performance of concrete.

i. ***Self-Compacting Concrete***

Self-compacting concrete (SCC), also called self-consolidating concrete, is capable of flowing and compacting under its own weight. At the same time, it is cohesive enough to fill spaces of almost any size and shape without segregation. This makes SCC especially useful where placement is difficult, such as in heavily reinforced concrete or in complex formwork. This technology, which was developed in Japan in the 1980s, is based on increasing the amount of fines, such as fly ash or limestone fillers, without changing the moisture content compared to ordinary concrete. This changes the rheological behavior of the concrete.

j. ***Concrete with Reactive Powder***

Reactive powder concrete (RPC) was first patented by a French construction company in 1994. The properties of RPC are achieved by: (1) eliminating coarse aggregates, where only very fine powders are used, such as crushed quartz sand and silica fume; (2) optimizing the grain size distribution to compact the mixture; (3) post-setting heat treatment to improve the microstructure; (4) addition of steel and synthetic fiber (about 2% by volume); and (5) the use of a superplasticizer to lower the w/c ratio, which is usually less than 0.2 to simultaneously improve the rheology of the paste.

References

1. Nawy, E.G.: Prestress Concrete; A Fundamental Approach, 4th edn. Pearson Education Inc. (2003)
2. Neville, A.M.: Properties of Concrete, 5th edn. Pearson Education Limited Edinburgh Gate Harlow Essex CM20 2JE, England (1998)
3. Krishna, Civilread.: CIVIL READ Copyright © 2022. https://civilread.com/contact-us/
4. ACI-318–19: Building Code Requirements and Commentary. https://www.concrete.org/tools/318buildingcodeportal.aspx.aspx. (2014)
5. Kosmatka, S., Kerkhoff, B., Panerese, W.: Design and Control of Concrete Mixtures, 14 edn. Portland Cement Association, Skokie, Ill (2002)
6. Kosmatka, S.H., Panarese, W.C., Kerkhoff, B.: Design and control of concrete mixtures. In: Chapter 17: High-Performance concrete, Portland Cement Association, 13th edn. http://www.ce.memphis.edu/1101/notes/concrete/PCA_manual/Chap17.pdf. (1988)
7. Indeed: Types of cement and their common uses in construction. https://www.indeed.com/career-advice/career-development/cement-types (2022). Accessed 3 November
8. Hanson: Types of Cement Used In The Construction Industry. https://www.hanson.my/en/types-cement-construction-industry (2020)
9. Tyupinaa, E.A., Sazonov, A.B., Sergeecheva, Y.V., Shestakov, I.A., Tuchkova, A.I., Nikitin, A.V.: Application of thermally expanded graphite for the cementation of cesium-and tritium-containing waste oils. Inorg. Mater. Appl. Res. **7**(2), 196–203 (2016)
10. Suparjo, A.Y.P.: Prediction of moment strength of steel reinforcement concrete bar in NaCl environment. Int. J. Civil Eng. **9**(11), 50–55 (2022)
11. Turner, F.: Federal highway administration research and technology. Report Portland Cement Concrete Pavements Research, 200 New Jersey, Washington, DC. https://www.fhwa.dot.gov/publications/research/infrastructure/pavements/pccp/thermal.cfm (2016)
12. ACI-222: Guide to Design and Construction Practices to Mitigate Corrosion of Reinforcement in Concrete Structures (2001)
13. Saad, M., Abo-El-Enein, A.A., Hanna, G.B., Kotkata, M.F.: Effect of temperature on physical and mechanical properties of concrete containing silica fume. Cem Concr Res. **26**(5), 669–675 (1996). https://doi.org/10.1016/S0008-8846(96)85002-2
14. Lin, W.-M., Lin, T.D., Powers-Couche, L.J.: Microstructures of fire-damaged concrete. Mater. J. **93**(3) (1996)
15. Li, X.J., Li, Z.J., Onofrei, M., Ballivy, G., Khayat, K.H.: Microstructural characteristics of HPC under different thermo-mechanical and thermo-hydraulic conditions. Mater. Struct. **32**, 727–733 (1999)
16. Consolazio, G.R., McVay, M.C., Rish, J.W.: Measurement and prediction of pore pressures in saturated cement mortar subjected to radiant heating. ACI Mater. J. **95**, 5 (1998). https://hal.archives-ouvertes.fr/hal-02566936/document
17. Ozawa, M., Uchida, S., Kamada, T., Morimoto, H.: Study of mechanisms of explosive spalling in high-strength concrete at high temperatures using acoustic emission. Constr. Build. Mater. **37**, 621–628 (2012)
18. Coniglio, N., Spitz, N., Mansori, E.: Designing metallic surfaces in contact with hardening fresh concrete: a review. Constr. Build. Mater. **255**, 119384 (2020)
19. Mlodecki, J.: Adhesion forces of polymer modified concrete and plain concrete to steel in molds and in reinforced concretes. In: Adhesion Between Polymers and Concrete, pp. 55–63. Springer, Boston, MA (1986)
20. Mazkewitsch, A., Jaworski, A.: The adhesion between concrete and formwork. In: Adhesion Between Polymers and Concrete, pp. 67–72. Springer, Boston, MA (1986)
21. ASCE: High-performance construction materials and systems. Technical Report 93–5011, American Society of Civil Engineers, New York (1993)
22. Russell, H.G.: ACI Defines High-Performance Concrete. Concrete International, American Concrete Institute, Farmington Hills, Michigan, pp. 56–57 (1999)

23. Callister, W.D., Rethwisc, G.D.: Materials Science and Engineering an introduction. Department of Metallurgical Engineering, The University of Utah. Rethwisch, G.D., Department of Chemical and Biochemical Engineering, The University of Iowa (2017)
24. Bickley, J.A., Fung, R.: Optimizing the economics of high-performance concrete. In: A Concrete Canada and Canadian Cement Industry Joint Research Project. Cement Association of Canada, Ottawa, Ontario, p. 21 (2001)
25. William, C., Kosmatka, Steven H., Panarese.: Design and Control of Concrete Mixtures. Chapter 17: High-Performance concrete, Portland Cement Association, Thirteenth Edition, 1988. http://www.ce.memphis.edu/1101/notes/concrete/PCA_manual/Chap17.pdf
26. Dunaszegi, L.: HPC for Durability of the Confederation Bridge. HPC Bridge Views, Federal Highway Administration and National Concrete Bridge Council, Portland Cement Association, September/October, p. 2 (1999)

Chapter 3
Types and Causes of Concrete Damage

Abstract This chapter highlights the susceptibility of concrete structures to various forms of damages. Chemical attack poses a common threat, with acids, chlorides, salts, and alkalines being significant factors. Furthermore, concrete deterioration is contributed to by sulfate attack and carbonation, while alkali-aggregate reaction and alkali-carbonate reactivity also hold potential for damage. External factors play a significant role in concrete damage as well. Over time, abrasion, erosion, and erosion cavitation gradually degrade the concrete surface. Exposure to fire or prolonged high temperatures compromises the structural integrity of concrete. Volume changes, such as shrinkage or expansion, can result in thermal cracking. Additionally, surface defects impact both the aesthetics and performance of concrete structures. To ensure the long-lasting durability of concrete structures, it is crucial to identify and address these various types and causes of damage. The implementation of proper maintenance, timely repairs, and preventive measures plays a pivotal role in mitigating these issues and extending the lifespan of concrete infrastructure. By effectively managing these risks, the integrity and longevity of concrete structures can be preserved.

3.1 Damage of Concrete

Concrete is a material which is resistant to environments that contain water, air and chemicals. As long as there is no physical damage, the age of concrete can reach hundreds of years. These properties make concrete suitable for structural buildings built in seawater, corrosive atmosphere, acidic soil, and buildings in industrial environments. However, if the quality of the concrete is poor, the concrete may tend to damage. Damage of concrete is usually preceded by poor surface quality. The damaging effects of chemicals on concrete are caused by the reaction of concrete with chemical elements from outside that enter through the cracks and porosity the concrete. The effort should be achieved that the concrete is always in a solid state, no cracks and low porosity so that active elements cannot be penetrated. If the chemical impurities reach out the concrete, the number of impurities must be kept below the minimum concentration unless the steel will experience a corrosion reaction.

Y. P. Asmara, *Concrete Reinforcement Degradation and Rehabilitation*,
Engineering Materials, https://doi.org/10.1007/978-981-99-5933-4_3

Table 3.1 Costs due to corrosion in various infrastructure in the US [3]

Field	Annual fee ($B)
Bridges and highways	8.3
Gas and liquid pipelines	7
Waterways and harbours	0.3
Storage of hazardous materials	7
Gas distribution	5
Drinking water and waste disposal	36.6

The impact caused by the corrosion reaction in reinforcing steel is very deterious where a number of infrastructures can be damaged and a number of facilities are not functioning. In addition, indirect corrosion impact are environmental pollution. Corrosion damage to the infrastructure has been costly. Costs incurred due to corrosion consist of direct costs which are repair costs if damage occurs and indirect costs. Indirect costs are a number of costs that must be paid for corrosion prevention activities for example: surface coating, cathodic protection, training management, warehousing and implementation of corrosion control systems in infrastructure [1, 2] . In the US, it was noted that the costs incurred due to damage to various sectors are shown in Table 3.1.

Concrete buildings that have been constructed for a long-time experience various process of destruction. Damage to concrete can occur due to an imperfect manufacturing that triggers further damage. The surrounding chemicals from atmosphere may interact with the concrete compositions, which also results in damage. The following are the types of processes that can damage reinforced concrete.

3.2 Chemical Attack

3.2.1 Acids

In general, portland cement for concrete does not have good resistance to acidic materials. Regardless of its composition, concrete will not last long if it is exposed to solutions with a pH of 3 or lower. However, some of weak acids may be accepatble, especially if exposure is not to long. The acid is capable in reacting with the calcium hydroxide of hydrated portland cement. In most cases, the chemical reaction forms water-soluble calcium compounds [4]. Combustion process from fuels that produce sulfur gas can combine with water vapor to form sulfuric acid. Sulfuric acid can also come from certain bacteria in the soil that turn sewage into sulfuric acid. Sulfuric acid is very aggressive to concrete because the calcium sulfate formed from the acid reaction will also damage concrete through sulfate attack.

In addition, organic and mineral acids that can attack concrete are acid-containing or acid-producing substances, such as acid industrial waste, silage, fruit juices, and

Table 3.2 Types of chemical elements that can damage concrete [5]

Aluminum chloride	Mustard oil*
Aluminum sulfate*	Nitric acid (all concentrations)
Ammonium bisulfate	Perchloric acid, 10%
Ammonium nitrate	Potassium dichromate
Ammonium sulfate*	Potassium hydroxide (>25%)
Ammonium sulfide	Promote moderate deterioration of concrete
Ammonium sulfite	Promote rapid deterioration of concrete
Ammonium superphosphate	Rapeseed oil*
Ammonium thiosulfate	Slaughterhouse waste2
Calcium bisulfite	Sodium bisulfate
Castor oil	Sodium bisulfite
Chemicals that deteriorate concrete	Sodium hydroxide (>20%)
Cocoa bean oil*	Sulfite liquor cocoa butter*
Coconut oil*	Sulfuric acid, 10–80*
Cottonseed oil*	Sulfuric acid, 80% Oleum*
Fish liquor1*	Sulfurous acid
Hydrochloric acid (all concentrations)*	Tanning liquor (if acid)
Hydrofluoric acid (all concentrations)	Zinc refining solutions3

* *Used in the food or beverage processing industry*

sour milk, will also cause damage. Animal manure contains substances which can be oxidized in the air to form acids which attack concrete. Acid rain which has a pH of 4 to 4.5 can scratch concrete surfaces. Any water that contains bicarbonate ions also contains free carbon dioxide which can partially dissolve the calcium carbonate. Water with aggressive carbon dioxide (CO_2) corrodes concrete through an acid reaction and can attack concrete and portland cement products. Acidic soil absorbs calcium and can attack concrete, especially in porous concrete. Chemicals can attack concrete by dissolving calcium in the cement paste to form silica gel. To prevent damage due to acid attack, Portland cement concrete generally must be protected from an acidic environment by a protective layer on the surface of the concrete (see Table 3.2).

3.2.2 Chloride

Reinforced concrete exposed to an environment containing chloride ions can be damaged when reinforcement of steels corrode (see Figs. 3.1 and 3.2). Chloride ions from seawater can reach to reinforced concrete and cause corrosion if oxygen and water vapor are also present. Chloride dissolved in water penetrate through

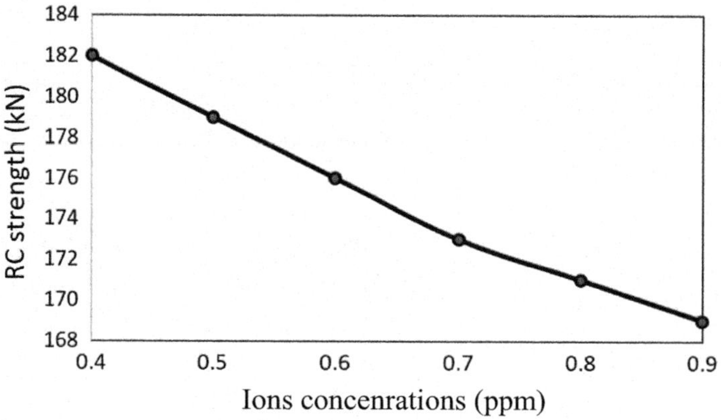

Fig. 3.1 Effect of chloride concentration in concrete-on-concrete strength [6]

concrete and reach steel if there are cracks. Chloride causes corrosion by destroying the protective oxide film formed on steel surface. The risk of corrosion increases with increasing the chloride content in the concrete. When chloride content on the steel surface exceeds a certain limit, corrosion will start to occur. However, only the chlorides in certaian amount which dissolve in water cause corrosion. Tables 3.3 and 3.4 below shows maxium alloable chloride contents allowed in the concrete and threshold of chlorine content in various types of steel in OPC concrete.

3.2.3 Salt and Alkali

The chlorides, nitrates, ammonium, magnesium, aluminum and iron all damage concrete where ammonium is the most damaging element. Most of the ammonium salts are destructive because in the alkaline environment of concrete they release ammonia gas and hydrogen ions. The release of hydrogen is followed by dissolving the calcium hydroxide from the concrete. Strong alkalis (more than 20%) can also cause concrete disintegration [5].

3.2.4 Sulfate Attack

Sodium, potassium, calcium, or magnesium sulfates either exist in the soil or dissolved in groundwater. Sulfates can attack concrete by reacting with hydrated cement compounds. These reactions can cause stresses which is sufficient to displace the cement and result in a loss of cohesive forces. Sodium sulfate reacts with calcium

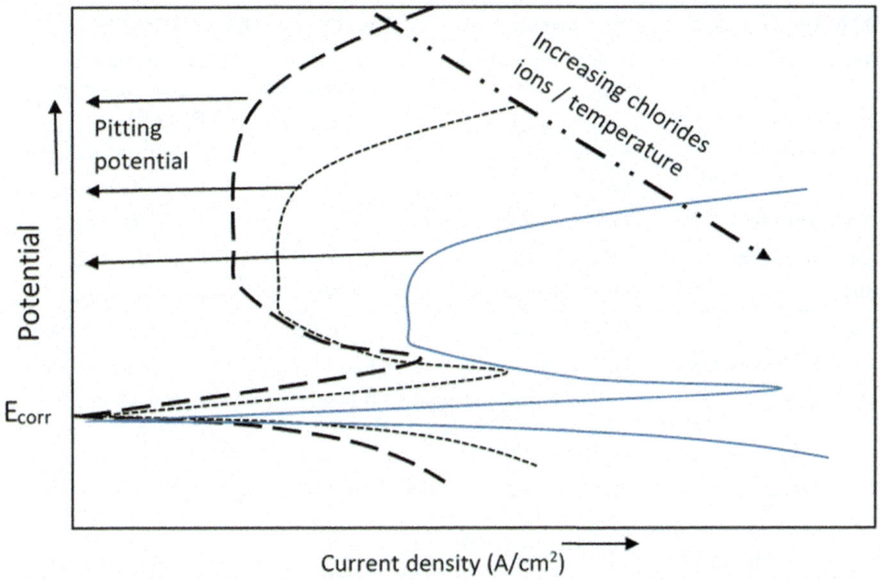

Fig. 3.2 Schematicallly effect of chloride levels and temperature on the polarization diagram. Either chloride or temperature causes the passive layer to thin and reduces the potential for pitting corrosion [7]

Table 3.3 The maximum permissible chloride ion content in reinforced concrete [8]

Concrete type	Maximum cl-*
Pre-stressed concrete	0.06
Reinforced concrete is exposed to chlorides directly	0.15
Reinforced concrete dry or protected from moisture	1.00
Other reinforced concrete construction	0.30

* *Water-soluble chloride, percent by weight of cement*

hydroxide and hydrated calcium aluminate to form ettringite and gypsum. Magnesium sulfate also attacks concrete by forming ettringite, gypsum, and brucite (magnesium hydroxide). Brucite forms primarily on the concrete surface, consumes calcium hydroxide, lowers the pH, and then decomposes the hydrate calcium silicate. This attack will occur more if there is a wet/dry cycle. As the water evaporates, sulfates can accumulate on the surface of the concrete to increase concentration which is potentially to cause deterioration. Porous concrete is susceptible to concrete weathering as effect of salt crystallization. Examples of salts known to affect weathering of concrete are sodium carbonate and sodium sulfate. As the concrete dries, the salt solution can rise to the surface via capillary reactions. As a result of evaporation, the solution becomes saturated and crystallization of the salt occurs. Formation of salts

Table 3.4 Threshold of chlorine content in various types of steel in OPC concrete [9]

Metal type	Concrete conditions	Threshold concentration	Investigation of corrosive specimen methods
Carbon steel	Mortar suspension OPC	2.4%	Anodic polarization
SS	OPC solution	4%	Corrosion rate
Alloy 254 SMO	OPC solution	10%	Corrosion rate
Galvanized Zinc	OPC solution	kg/m3	Corrosion rate
304 L SS	OPC solution	4%(W)	Anodic polarization
316 L SS	OPC solution	5.5%(W)	Anodic polarization
LDX 2101 Duplex SS	OPC solution	6.5–7% (W)	Anodic polarization
Carbon steel	Concrete slab with added Clto various exposure conditions OPC	0.1–0.19	Corrosion rate, visual inspection, mass loss
Carbon steel	Cement with high or low alkali content; Mortar 80% RH 100% RH	0.6–1.8 0.5–1.7	Corrosion rate

can supress the concrete lead to cracks of the concrete. As a result of evaporation, the solution becomes saturated and crystallization of the salt occurs [10].

Seawater also contains sulfates but the effects are not as severe as sulfates in groundwater. Sulfate resistance is achieved when concrete made of low water-cement ratio (w/c) or cement with limited amounts of tricalcium aluminate. To reduce the risk of sulfate attack, concrete uses cement that meets the requirements as *ASTM C 1157*, namely cement type of MS (moderate sulfate resistance) and cement type of HS (high sulfate resistance) (*ASTM C 595*). Studies have shown that some pozzolans and blast-furnace slag increase the resistance of sulfate-exposed concrete. The use of fly ash meets the requirements as *ASTM C 618* class F. Some pozzolans, especially some class C fly ash reduce sulfate resistance. Therefore, pozzolans selected to increase sulfate resistance must be tested to confirm their behavior. Calcium chloride reduces sulfate resistance, so it should not be used as cement mixture.

3.2.5 *Carbonation*

Carbonation occurs when carbon dioxide from the air penetrates concrete and reacts with hydroxides, such as calcium hydroxide to form carbonates. In the reaction with calcium hydroxide, calcium carbonate is formed [11]:

$$Ca(OH)_2 + CO_2 \rightarrow CaCO_3 + H_2O \tag{3.1}$$

Table 3.5 Criteria for damage to concrete based on the carbonation reaction. *dcar/dcoat is depthcarbonation and concrete cover thickness [13]

dcarb/d*coat	Concrete condition	Rebar condition	Risk of corrosion
<0.5	No cracking	Passive	Negligible
>0.5	No cracking	Passive	Low
≈1.0	Small cracking	Low to moderate corrosion	High
>1.0	Cracks, minor detachment/ spalling	Moderate to high corrosion	Very High
>>1.0	Cracks, high detachment/ spalling	High corrosion with substantial loss of section	Very High/Severe

This reaction causes the pH to be 8.5 which is at this level the passive film on the steel is unstable. Carbonation is generally a slow process. In high strength concrete, it is estimated that carbonation will occur at rates of up to 1.0 mm per year. The amount of carbonation increases significantly in high w/c concrete. Carbonation is highly dependent on the relative humidity of the concrete. The highest degree of carbonation occurs when the relative humidity is between 50 and 75%. Under 25% Relative Humidity the degree of carbonation has no effect. Above 75% relative humidity, water vapor in the cement pores can resist CO_2 penetration [4]. Corrosion due to carbonation often occurs in areas of buildings exposed to rainfall and protected from sunlight.

Carbonation of concrete also reduces the amount of chloride ions required for corrosion reactions to occur. In fresh concrete with a pH of 12–13, about 7.000 to 8.000 ppm of chloride is required to start corrosion of the steel in the concrete. However, if the pH decrease to 10 to 11, the chloride ion limit for initiating the corrosion reaction is much lower, which is below 100 ppm [12]. The risk of corrosion due to carbonation of concrete depends on the depth of carbonation in relation to the thickness of the concrete cover shown in Table 3.5.

3.2.6 Alkali-Aggregate Reaction

In most concrete, aggregate is chemically inert. However, some aggregates can react with the alkaline hydroxide in the concrete causing expansion and causing cracks on the surface of the concrete. The alkali-aggregate reactivity takes two forms: the alkali-silica reaction (*ASR)* and the alkali-carbonate reaction (ACR). This destructive chemical reaction changes the composition of the aggregate by forming a soluble and viscous sodium silicate gel ($Na_2SiO_3\ nH_2O$, *also noted* $Na_2H_2SiO_4\ nH_2O$, *or NSH (sodium silicate hydrate)*), depending on the adopted convention. This hygroscopic gel swells and increases in volume when it absorbs water. It can exert expansive stresses within siliceous aggregates and cause peeling and loss of concrete strength, eventually leading to failure.

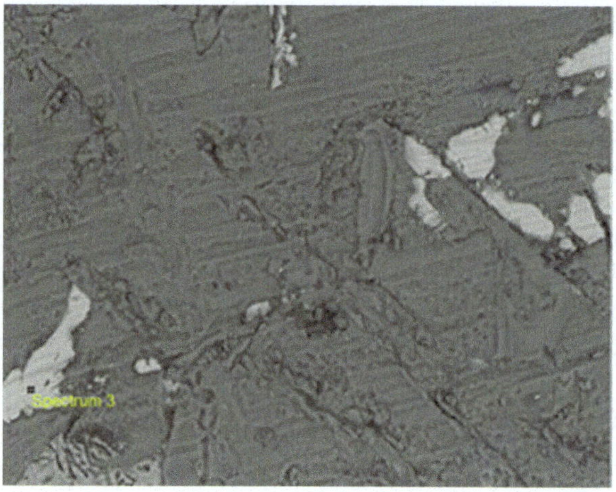

Fig. 3.3 Schematic of ASR attack on concrete [14]

From Fig. 3.3 above, an indicator of ASR is the formation of random deep crack patterns which cause concrete surface to peel off. Cracks usually appear in areas with access to moisture, such as locations close to the waterline in pillars, behind retaining walls, near joints and free edges of pavements, or columns subjected to axial loads. Certain amount of moisture concentration is required to produce expansion which give detrimental effects to the concrete. Alkali-silica reactivity can be reduced significantly by maintaining the concrete in dry condition. Reactivity can almost be stopped if the internal relative humidity of the concrete is set below 80%. Warm seawater and dissolved alkalis can intensify the alkali-silica reactivity. Alkali-silica reactivity can be controlled using a mixture of silica fume minerals, fly ash, and ground-granulated blast furnace slag. Class F fly ash can also reduce reactivity by up to 70%. In some cases, lithium compounds have been shown to be effective at reducing ASR.

The mechanism of ASR causing concrete deterioration can be explained in four steps as follows: Highly alkaline solutions (NaOH/KOH) attack the silica aggregates (dissolving silicic acid at high pH) and convert the non-crystallized or amorphous silica into a soluble but highly alkaline silicate gel viscous (NSH, KSH). Consumption of NaOH/KOH by dissolving reaction of amorphous silica lowers the pH in the pores of the cement paste. This allows the dissolution of $Ca(OH)_2$ (portland) and increases the concentration of Ca^{2+} ions into the cement pore water. The calcium ions then react with the dissolved sodium silicate gel to convert it into solid calcium silicate hydrate (CSH). CSH forms a continuous poor permeable layer on the outer surface of the aggregate [15]. Stages of expansion in concrete due to ASR attack [11].

a. Alkalies + Reactive Silica → Gel
b. *Gel + Moisture → Expansion.*

3.2.7 *Alkali-Carbonate Reactivity*

Certain dolomite rock reactions are associated with the alkali-carbonate reaction (ACR). Dedolomitization, or destruction of dolomite, is usually associated with expansive alkali-carbonate reactivity. The ACR reaction and subsequent crystallization of brucite can cause a considerable expansion. The losses caused by the alkali-carbonate reaction are like those caused by the alkali-silica reaction, however, the alkalicarbonate reaction is relatively rare because aggregates susceptible to this reaction are usually unsuitable for application in concrete for reasons of strength.

3.3 External Factors Causing Damage to Concrete

3.3.1 *Abrasion/Erosion*

Abrasion damage occurs when the concrete surface is unable to withstand loads in the form of friction and impact so that the fine and coarse aggregates are exposed. Wind can also cause abrasion by carrying fine particles. Abrasion often occurs in the concrete floors of highways, dams, waterways, offshore structures and tunnels. Many floors of highways, especially in industrial environments, experience abrasion by heavy vehicles and transportation. Sand can be used to increase abrasion resistance (traction) with the condition that special aggregates are hard or tough must be used. Compressive strength is the most important factor against abrasion force. In water structures, the causes of abrasion are mud, sand, gravel, stones, and other water-borne debris hitting the concrete surface.

While high-quality concrete can withstand the high velocity of water for years with little or no damage, it is not resistant to the abrasive force from debris or repeated impacts on concrete surface. Abrasion erosion is identified by its appearance, which is shown by the presence of wear following the erosion direction and the holes formed by cavitation erosion. As the case with traffic wear and tear, abrasion damage to hydraulic structures can be reduced by using strong concrete with hard aggregate.

3.3.2 *Erosion Cavitation*

Cavitation erosion is the damaging attack on the surface of a material by gas or vapor bubbles. The sudden burst of air bubbles causes a sudden change in pressure resulting in a microjet force hitting the surface of the object. Cavitation generates deformation, pitting, noise and vibration. If cavitation occurs together with erosion, significant damage to the concrete will occur. After cavitation damage occurs, it is followed by changes in flow patterns that add to the damage mechanism. Cavitation is recognized by the shape of the attack. Cavitation produces wear on the coarse aggregate particles and has irregular and rough edges. Severe cavitation damage will usually form a "*Christmas tree*" configuration on the surface of the waterways. Although proper material selection can increase concrete's cavitation resistance, the only effective solution is to design hydraulic structures with high strength concrete to reduce or eliminate cavitation-inducing factors.

3.3.3 Fire/Heat

Concrete can be used in construction over a wide range of normal temperatures. But when exposed to fire or very high temperatures, concrete can lose its strength and stiffness. Figure 3.4 shows the effect of high temperature on the compressive strength, flexural strength, and elastic modulus of concrete [16]. As shown in the Fig. 3.4, the modulus of elasticity is the most sensitive to high temperature, followed by flexural strength and compressive strength. A number of studies have found trends in the general condition of a concrete exposed to high temperatures as follows:

- Concrete that undergoes thermal cycling suffers a greater loss of strength than concrete that is at constant temperature. This is due to the unequal expansion coefficient between cement paste, steel and aggregate.
- Concrete containing limestone and calcium aggregates has better thermal resistance properties than concrete containing silica aggregates [17]. One study showed no difference in the performance of dolostone and limestone [18]. Another study showed the performance of aggregates if sorted starting from the best, namely: refractory bricks, sedimentary stone, limestone, gravel, sandstone, and slag (terag).
- The loss in strength is independent of the compressive strength of the concrete.

Fig. 3.4 Various surface defects that can occur on the concrete surface due to the reaction of concrete exposed to the atmosphere

- Concrete with a higher aggregate-cement ratio has a slight decreasing in compressive strength, but it is contradictory for the modulus of elasticity. The lower the water-cement ratio (W/C), the less the loss of elastic modulus.
- If the residual water in the concrete is not permissible to evaporate, the compressive strength will greatly reduce. If heated too quickly, the concrete will flake as moisture tries to escape.

3.3.4 *Volume Change*

Concrete will change its volume for various reasons. The most common causes are fluctuations in moisture content and temperature. Restraint of volume changes, especially contraction, can cause cracking if the volume changes result tensile stress exceeds the tensile strength of the concrete. Shrinkage cracks occur when water evaporates from the surface of freshly poured concrete faster than it is replaced by seepage water. As confinement occurs beneath the curing surface layer lead tensile stresses to develop in the concrete. It produces shallow cracks of varying depth. These cracks are often quite wide on the surface. Cracks due to plastic shrinkage can be prevented by reduce the rapid loss of water from the surface of the concrete. Mist spraying, plastic sheeting, windbreaks, are usually used to prevent over evaporation.

3.3.5 *Thermal Cracking*

Thermal cracking in concrete occurs because concrete expands when heated and contracts when cooled. The thermal expansion and contraction of concrete varies depending on the type of aggregate, cement content, water/cement ratio, age of concrete, and relative humidity. Aggregates are materials that have the greatest influence on thermal cracking. The concrete design must be designed to provide room for expansion and shrinkage. Excessive damage may occur during construction when the concrete has not attained the design strength. Thermal cracking can also occur when concrete is subjected to a load exceeding the design load.

3.3.6 *Surface Defects*

Various defects can occur on concrete surfaces when concrete is making or after concrete is in service. Defects can be avoided by using the right materials and adequate construction practices. Various defects such as air voids that form on the surface are the result of air bubbles trapped on the concrete surface which are formed during casting and consolidation. Pits on vertical surfaces occur due to excessive fine aggregate content, entrapped air, or both (see Fig. 3.4). Using a vibrator with a high amplitude can also cause imperfections in the concrete surface. The following are the types of defects that commonly occur on concrete surfaces:

- *Honeycomb* occurs, it is when the mortar fails to fill the spaces between the coarse aggregates. Overly dense reinforcement, segregation and insufficient fine aggregate content can contribute to honeycomb defects. This defect can be avoided by using higher concrete vibrations or increasing the flowability of the concrete (see Fig. 3.5).
- *Form tie holes.*The form of this defect is a cavity that is deliberately made in the surface of the concrete which is useful for preventing leakage or adding a concrete cover over the reinforcing steel (see Fig. 3.6).
- *Cold joints.* Cold joints are discontinuities in concrete components resulting from delays in the installation of the concrete being joined. Because the cooling speed is too fast so that the unification of the concrete is not achieved. Cold joints produce

Fig. 3.5 Types of honeycomb damage on concrete surfaces

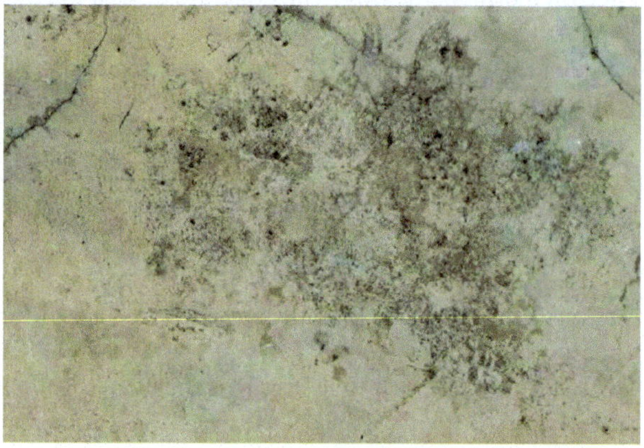

Fig. 3.6 Tie hole defects

Fig. 3.7 Streak defects (stripes)

defects in the form of visible lines of joints, where one layer of concrete has hardened before the next concrete.

- *Form streaks.* Streaks formed due to mold. Excessive vibration and use of concrete that is too wet or concrete with high shrinkage increases the chances of scratch formation. It can also occur due to mold joints that are not tightly closed (see Fig. 3.7).
- Offset forms. Form of surface irregularities caused by molds that are not strong and experience shifts.
- Delaminations. Delamination occurs when air and water are trapped beneath the surface of the mortar. Air and water trapped on the surface of the concrete (see Fig. 3.8). Concrete delamination refers to the separation or detachment of layers within a concrete structure. It occurs when the bond between the layers of concrete weakens or fails, leading to the formation of hollow areas or pockets between them. This can result in visible cracks, spalling, or a hollow sound when the delaminated area is tapped.
- *Dusting.* Is the formation of dust on concrete that easily peels off from the hardened concrete surface. These defects are formed due to a thin and weak surface layer consisting of water, cement, and fine particles. Wastewater that floats and flows back onto the concrete surface can cause dust. Another reason is using a mixture that is too wet, spreading dry cement over the surface to speed up drying. Heaters without ventilation can also cause dust due to the presence of carbon dioxide which reacts with the calcium hydroxide in the concrete and forms a weak layer of calcium carbonate on the surface (see Fig. 3.9).

Fig. 3.8 Delamination defects

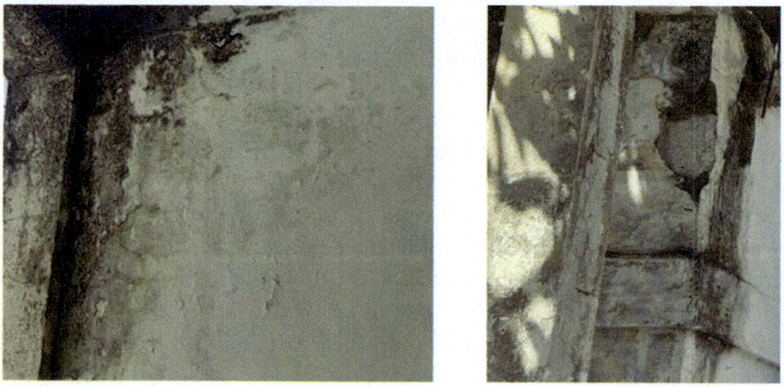

Fig. 3.9 Dusty concrete surface

- *Popouts.* Popout crack in concrete is a portion of the concrete surface breaks or pops out, typically leaving behind a void or depression. It occurs when the aggregate particles near the surface of the concrete experience internal pressure and dislodge from the matrix, causing a visible crack and the displacement of material. The aggregate absorbs moisture at a given humidity which creates sufficient internal pressure to damage the concrete surface. Popout cracks are commonly caused by various factors, including the presence of reactive aggregate, moisture changes, freezing and thawing cycles, and internal stresses within the concrete. They can affect durability of concrete which depends on the severity and location of the cracks (see Fig. 3.10).
- *Subsidence cracks.* Subsidence cracks occur due to reinforcement in the concrete. These cracks occur when the concrete dries where there is a decrease in the level of cement adhering to the reinforcement or molding tools (see Fig. 3.11).

Fig. 3.10 Popout crack [19]

Fig. 3.11 Cracks caused by shifting reinforcement

- *Crazing* which is a fine crack pattern that does not penetrate deep below the surface and is usually only a surface problem. This type of defect is almost invisible, except when the concrete dries after the surface has been wet. Preventing excessive evaporation during placement and proper curing can prevent crazing (see Fig. 3.12).
- *Spalling*. Spalled concrete can be seen from the cracked and delaminated concrete surface. This fraction is not caused by the influence of loading or collision. The formation of spalling is caused by, among other things: corrosion of the reinforcement or sulfate attack on the concrete so that the concrete layer becomes brittle (see Fig. 3.13).

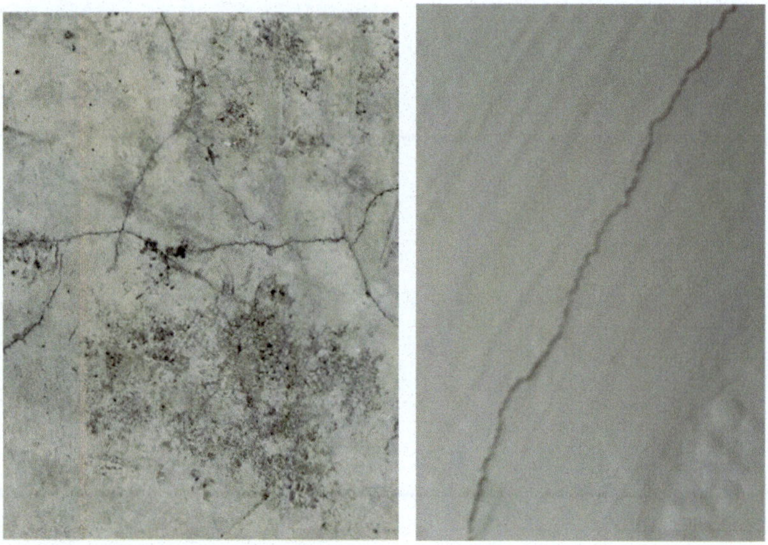

Fig. 3.12 Crazing crack

Fig. 3.13 Spalling damage

- *Blistering*. Blistering damage forms blisters on the concrete surface. These blisters are formed during the concrete manufacturing process where there has been rapid drying on the surface before the inside. As a result, air bubbles or water are trapped under the surface. Other causes of blistering are: too much or too little vibration is used, lack of exhaust accessair, too much air around the concrete manufacturing site, the foundation has a colder temperature than concrete, the size of the concrete is thick so it takes a long time for air and water to flow up to the surface (see Fig. 3.14).

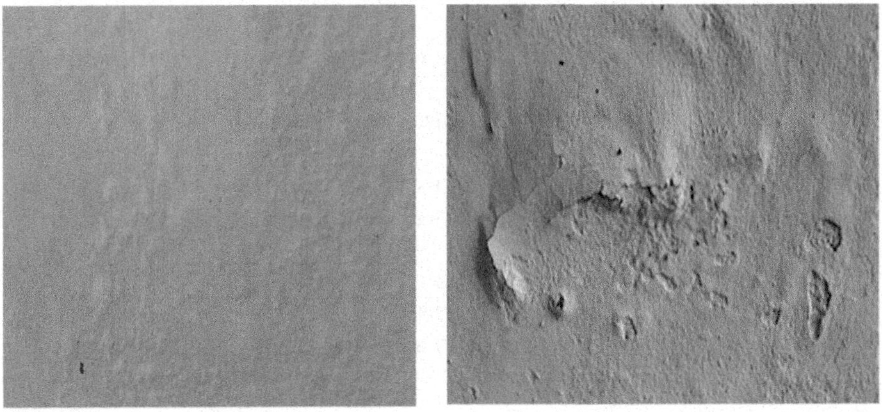

Fig. 3.14 Blistering defects

Corrosion damage can be associated with electrical processes. So that damage to reinforced concrete can also be known indirectly by knowing the magnitude of the potential, resistance, current density. Criteria for damage based on the value of resistance, potential and current density are shown in Tables 3.6 and 3.7 as follows.

Table 3.6 Reinforced concrete damage criteria [13]. **a** Criteria based on the corrosion potential of reinforcing steel. **b** Criteria based on the value of corrosion resistance of reinforcing steel. c Criteria based on the corrosion current of reinforcing steel

Risk of corrosion	OCP values	
	mV versus Sce	mV versus CSE
Severe	<−426	< −500
High	<−276	< −350
Intermediates	−126 to −275	−350 to −200
Low	> −125	> −200
Risk of corrosion	ρ/kΩm	
Insignificant	>1	
Low	0.2 to 1	
High	0.1 to 0.2	
Very high	<0.1	
Construction steel conditions	icorr/μA cm $^{-2}$	
Passive	<0.1	
Low to moderate corrosion	0.1 to 0.5	
Moderate to high corrosion	0.5 to 1.0	
Very high	>1.0	

Table 3.7 Types of cracks and their causes [20]

Crack type	Sighting	Reason	Time formed cracks
Plastic settlement	Above or in line there is a decrease in the reinforcement area	Imperfect mixing, bleeding and vibration	10 min-3 h
Plastic crack	Diagonally/randomly	Excessive evaporation	30 min-6 h
Expansion	Transverse (transverse)	Excessive temperature changes	1 day – 3 weeks
Cracks when drying	Patterned crack	Excess water, too wide joints	For weeks
Freezing/thawing	Align with concrete	Lack of water channels aggregate quality is not good	Week–month
Corrosion	On steel	Not enough cover, chlorides get into the concrete	More than 2 years
Alkaline aggregate reaction	Patterned, parallel to the restrained side	Aggregate reactive, alkaline, water vapor	More than 5 years
Sulfates	Patterned	There are sulfates	1–5 years

3.4 Concrete Treatment

Concrete treatment or rehabilitation is crucial practices to avoid early damages and deterioration. It varies depending on the nature, extent of the damage, specific circumstances, and severity of delamination. Here are some suggested actions that can be taken for concrete rehabilitation and improve quality of concrete structures.

- Remove and replace: In this method, the delaminated concrete is entirely removed, and new concrete is poured or applied in its place. This approach ensures a fresh bond and eliminates the weakened or damaged sections.
- Bonding agents: Bonding agents are specialized materials that improve the bond between existing concrete layers. These agents are applied to the delaminated area before new concrete is added, promoting adhesion and enhancing the bond strength.
- Epoxy injection: Epoxy injection involves injecting epoxy resin into the delaminated areas. The resin fills the voids and forms a strong bond, effectively restoring the integrity of the concrete. This method is particularly useful for small-scale delamination.
- Protective coatings: Applying protective coatings can help prevent further moisture infiltration and damage. These coatings create a barrier that seals the concrete surface, reducing the risk of delamination caused by water or chemical penetration.

- Strengthening techniques: Depending on the severity of delamination and the structural requirements, reinforcing methods can be employed. This may include adding additional steel reinforcement bars or fiber-reinforced polymers (FRP) to enhance the strength and stability of the affected area.
- Crack repair techniques: If the delamination has resulted in visible cracks, various crack repair techniques can be employed. This may involve filling the cracks with epoxy, polyurethane, or specialized crack repair materials to prevent further separation and restore the integrity of the concrete.
- Proper diagnosis and assessment of the delamination issue are key to implementing effective repairs. Corrosion Protection and Reinforcement.
- Implement corrosion protection measures to prevent or slow down the corrosion of embedded reinforcement, such as applying corrosion inhibitors, cathodic protection systems, or epoxy coatings.
- Install additional reinforcement, such as carbon fiber wraps or steel plates, to strengthen weakened or structurally compromised areas.

References

1. Uhlig, R.W.: Corrosion and Corrosion Control: An Introduction to Corrosion Science and Engineering, 3rd edn. Wiley, New York (1985)
2. Callister, D., Jr. David G., Rethwisch, G.D.: Materials Science and Engineering An Introduction. Department of Metallurgical Engineering, The University of Utah (2017)
3. Koch, G.H., Brongers, M.T.: Corrosion Cost and Prevention Strategies in the United States. Federal Highway Administration, Mc Lean (2002)
4. ACI-210R-93: American Concrete Institute, Farmington Hills, Michigan 1993 (2001)
5. ACI 515.1R-79: American Concrete Institute, Farmington Hills, Michigan, Carette, GG (1979)
6. Asmara, Y.P., Nor, M.I., Anwar, S.N.R., Sugiman, S.: Remaining strength prediction of reinforced steel bar concrete structure in seawater environment. AIP Conf. Proc. **2489**, 1 (2022)
7. Fontana, M.G.: Corrosion Engineering, 3rd edn. McGraw-Hill, New York (1986)
8. ACI-318–02: American Concrete Institute, Farmington Hills, Michigan (2002)
9. Panca Asmara, Y., Kurniawan, T., Edy Sutjipto, A.G., Jafar, J.: Application of plant extracts as green corrosion inhibitors for steel in concrete-a review Indonesian. J. Sci. Technol. **3**(2), 158–170 (2018)
10. Mehta, P.K.: Sulfate attack on concrete: separating myth from reality. In: Concrete International, Farmington Hills, Michigan, pp. 57–61 (2000)
11. PCA: Concrete Slab Surface Defects: Causes, Prevention, Repair, IS177. Portland Cement Association (2001)
12. Montani, R.: Concrete's Forgotten Enemy, Concrete Repair Digest, pp. 330–333 (1996).
13. Figueira, R.B.: Electrochemical sensors for monitoring the corrosion conditions of reinforced concrete structures: a review. Appl. Sci. MDPI https://www.mdpi.com/2076-3417/7/11/1157. (2017)
14. Fournier, B., Nkinamubanzi, P.C., Chevrier, R.: Comparative field and laboratory investigations on the use of supplementary cementing materials to control alkali-silica reaction in concrete. In: Proceedings of the Twelfth International Conference Alkali-Aggregate Reaction, January Project: Alkali-Silica Reaction in Concrete. Laval University, National Canadian Research Council (2004)
15. Ichikawa, T., Miura, M.: Modified model of alkali-silica reaction. Cem. Concr. Res. **37**(9), 1291–1297 (2007)

16. Lankard, D.R., Birkimer, D.L, Fondriest, F.F., Snyder, M.J.: The Effects of Moisture Content on the Constitution and Structural Properties of Portland Cement Concrete Exposed to Temperatures Up to 500°F. Battelle Memorial Institute, Columbus Laboratories (1968)
17. Abrams, M.S.: Compressive Strength of Concrete at Temperatures to 1,600F, RD016. Portland Cement Association (1973)
18. Carette, G.G., Painter, K.E., Malhotra, V.M.: Sustained High Temperature Effect of Concretes Made with Normal Portland Cement, Normal Portland Cement and Slag, or Normal Portland Cement and Fly Ash. Concrete International, pp. 41–51 (1982)
19. Ham, S.: Effect of mixing and placing in hot weather on hardened concrete properties. Int. J. Conc. Struct. Mater. **7**(2) (2013)
20. Saire, J.: Troubleshooting Flatwork Finishing Course: Repair And Rehabilitation Of Structures. University of South Florida, Manual for Repair and Rehabilitation of Structures (2018)

Chapter 4
Corrosion Theory

Abstract The corrosion of steel in concrete is a significant concern in infrastructure and construction industries. This chapter discusses the theories, processes, and effects associated with the corrosion of steel embedded in concrete structures. Understanding the major components of corrosion and the various corrosion types is crucial for comprehending the underlying mechanisms and developing effective corrosion management strategies. Corrosion theories, including corrosion thermodynamics and corrosion kinetics, provide insights into the fundamental principles governing the corrosion process. By examining the corrosion rate, influenced by factors such as potential electrode and environmental conditions, the severity of corrosion can be assessed. The detrimental effects of corrosion on steel in concrete, such as structural deterioration and loss of functionality, highlight the importance of implementing robust corrosion risk management practices. Thus, proper corrosion risk management techniques are essential for preserving the integrity, durability, and service life of concrete structures. This chapter lays the foundation for a comprehensive exploration of the corrosion of steel in concrete, examining its theories, processes, effects, and the strategies employed in corrosion risk management. By addressing the challenges posed by corrosion, stakeholders in the construction industry can ensure the longevity and safety of concrete structures.

4.1 Corrosion Process

Corrosion is the process of damaging materials, especially those made of metal due to reaction with their environment. Corrosion involves an electrochemical process in which an electron transfer reaction occurs which produces changing in chemical properties of a material. Corrosion is basically a natural process which can be considered as the reverse process of metal extraction [1, 2]. The extraction process is a metal purification process from mineral compounds (oxides, sulfides, carbonates). To change the form of metal from ore requires energy (heating and melting) which result in increasing the internal energy of the metal. This causes the metal process an unstable condition. When metals are exposed to an environment, especially a reactive environment, they tend to release those energy into a lower internal energy state and

Y. P. Asmara, *Concrete Reinforcement Degradation and Rehabilitation*,
Engineering Materials, https://doi.org/10.1007/978-981-99-5933-4_4

Earth minerals ⟶ *Mineral compounds* ⟶ *Refining* ⟶ *Equipment components metal* ⟶ *Corrosion* ⟶ *Earth minerals*

Fig. 4.1 **a** Hematite (Fe_2O_3) and **b** Magnetite (Fe_3O_4). *Sources* http://www.mindat.org/photo-65718.html, http://www.irocks.com/db_pics/pics/k326a.jpg

recombine to their original energy state. In Fig. 4.1, it is provided pictures of an iron ore in nature which is in the form of hematite and magnetite [1, 3]. Figure 4.2 shows a natural cyclus of metals starting form ore in the earth.

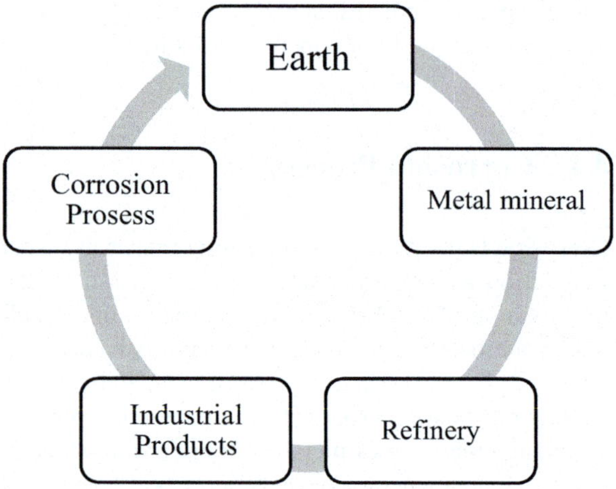

Fig. 4.2 The corrosion cycle in nature

4.1.1 Major Components of Corrosion

Corrosion will occur if there are four main factors present, namely:

- Presence of local anodes and cathodes either contained in the same material or with other materials. The anode and cathode create a potential difference.
- Active environment that can oxidize any of the metal parts. This environment is usually an electrolyte, which is a solution that can conduct electricity.
- The presence of a connection between the anode and cathode.

The anode is the location where metal dissolution will occur which is indicated by the release of electrons from the material. As a result, electrons flowing to other locations cause the metal to become ions which will dissolve into solution. If it is measured with a potentiometer, the anode has a lower reduction potential than the reduction potential of the cathode area.

The cathode is the place where electrons is consumed. As a result of the concentration of electrons in the cathode, the cathode will be negative and attract positively charged ions. The reduction potential at the cathode is higher than the reduction potential at the anode.

The process of producing (oxidation) and accepting electrons (reduction) is called an oxidation–reduction (redox) reaction. The redox reaction will take place in a solution that can conduct ions (electrolyte). However, at high pressures and temperatures that do not contain liquids, redox reactions can also take place which is then called dry corrosion.

Corrosion occurs when most or all of the atoms on the metal surface undergo oxidation. Over time, this process causes the metal structure to experience a thinning thickness and reduces the strength of the material. Most metals are easily oxidized and tend to lose electrons in both air and aqueous environments. When a metal loses electrons, at the same time oxygen gains electrons. This process takes place continuously to make the metal into metal oxide [1, 3].

The common form of corrosion that occurs in metals is uniform corrosion in which all metal surfaces have a uniform corrosion rate. Uniform corrosion is formed

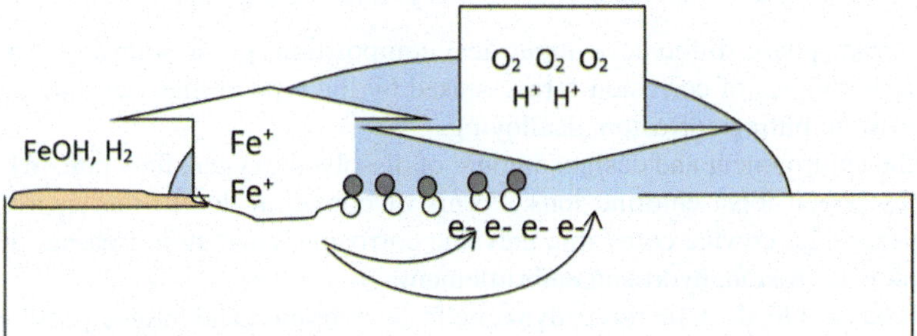

Fig. 4.3 Corrosion of ferrous metals in atmosphere condition. There is a flow of electrons from the anode to the cathode through the intermediary of water droplets or a thin film of water [1, 3, 4]

due to the presence of moisture (dew) attached to the metal surface. The presence of air pollution will accelerate the of corrosion process. Pollutants that are in the air are easily absorbed by water located on metal surfaces. If water is trapped between the two gaps, the corrosion process will take place more quickly. This type of corrosion due to crevices is called crevice corrosion (see Fig. 4.3).

4.2 Corrosion Effects

The effects of corrosion can be found directly or indirectly. Corrosion reduces the service life of metallic components and causes damage to objects, peeling paint, cracks and malfunctions of the production process. Corrosion is easily found on all equipment, for examples at: vehicle body panels, cooking equipment, water heaters, furnaces, stoves, washing machines, dryers and industrial facilities. Corrosion can cause various impacts ranging from impacts in the economic, health, environmental and industrial world. For example, the sudden collapse of infrastructure due to material depletion. Leaking gas pipes, contamination of canned food, environmental disasters caused by chemical tank explosions. Oil spills as a effect of local corrosion due to oil vessel leaks.

4.3 Corrosion Types

Corrosion reactions that occur in metals cause damage by producing various forms of corrosion. The form of the corrosion reaction results is influenced by the type of metal, environmental conditions, geometry, and metal structure. Knowing the type of corrosion that occurs in metals is useful for finding the causes of corrosion so that it does not happen again and again. Some forms of corrosion can be observed directly with the eye and some must use a microscope with a certain magnification [4].

The occurrence of this form of corrosion is influenced by the following factors:

- Material: phase difference, impurities, composition, grain boundary deterioration. Examples of corrosion forms caused by the type of material: intergranular corrosion, pitting corrosion, dealloying.
- Metal environment and design: amount of dissolved oxygen, flow rate, pH, levels of hydrogen ions, chlorine ions. Forms of corrosion due to the environment, for example: crevice corrosion, galvanic corrosion, cavitation, fretting, filiform, erosion corrosion, hydrogen embrittlement.
- Mechanical load: static force, dynamic force, vibration and fatigue stress. Examples of forms of corrosion due to mechanical loads are: stress corrosion cracking, fatigue corrosion, creep.

a. ***Uniform corrosion (uniform corrosion)***

Uniform corrosion usually occurs in carbon steel under atmospheric conditions. Uniform corrosion occurs because all metal surfaces are exposed to uniform conditions. This corrosion occurs on all surfaces at the same rate so it produce corrosion products with the same of thickness.

b. ***Pitting corrosion***

Pitting corrosion attacks materials that have the property of becoming passive by themselves or the material forms a passive layer due to the corrosion process as occurs in stainless steel, Aluminum alloy, Copper alloy. This type of corrosion attacks the metal at certain points and continuously to create deep holes.

c. ***Crevice corrosion***

Crevice corrosion occurs when there is a gap between the two material structures. Gaps in joints or deposits can be filled with water or mud. Metal construction joints such as nuts/bolts, welding, gaskets, rivets have the potential for crevice corrosion.

d. ***Galvanic corrosion***

When two different metals are connected to the electrolyte, there will be a potential difference between those two metals. Metals with a high oxidation potential will more easily release electrons to metals with a lower oxidation potential. Galvanic corrosion can also occur between different phases in an alloy.

e. ***Selective corrosion (dealloying, selective leaching)***

This corrosion occurs in one of the metal alloy components. Alloys are materials that consist of two or more elements. When one of the alloying elements becomes the anode to undergoes oxidation, at the same time the other components in the alloy become the cathode. This corrosion attack causes the alloy metal to experience a change in chemical properties followed by a decrease in strength.

f. ***Stress corrosion-Stress corrosion cracking (SCC)***

SCC destroys metal components rapidly due to the combination of tensile stress and corrosion occurring simultaneously. Metals may experience SCC along grain boundaries and in the presence of pre-existing cracks. There are several causes of SCC, namely expansion/contraction due to the cooling process, welding, residual stress, stress concentration or static load in the electrolyte environment.

g. ***Filiform corrosion***

Filiform corrosion is a type of localized corrosion that often occurs in aluminum and magnesium alloys coated with organic coatings. This type of corrosion also occurs on other plated metals such as zinc, iron and steel. Filiform corrosion tends to occur at high humidity, greater than about 75% and temperatures slightly above room temperature. Corrosion appears as threadlike filaments beneath the damaged coating.

h. ***Fatigue corrosion (CF)***

CF or fatigue corrosion is a metal crack caused by a combined reaction between fatigue stress and corrosion. Corrosion fatigue is similar to SCC which is degradation of materials due to stress and corrosion. The main difference with SCC is the character of the load appiled, SCC accepts static loads and CF accepts dynamic (repetitive) loads. Each metal has a fatigue life resistance (number of cycles until failure). In the presence of corrosive environmental medium, the fatigue life becomes shorter compared to fatigue in a dry air condition.

The presence of hydrogen ions causes faster damage, especially at higher temperatures. Corrosion fatigue caused by hydrogen ions can be prevented by selecting the right materials and keeping the environment aqueous at neutral or basic pH ($pH > 6$).

4.4 Corrosion Theory

4.4.1 Corrosion Thermodynamics

Free energy (G) is a thermodynamic quantity possessed by each element which depends on pressure and temperature. When an element undergoes a chemical reaction, the amount of free energy before and after the reaction will be either higher or lower. Calculating free energy can be used to determine the probability that a chemical reaction will take place spontaneously (naturally) or not. Theories thermodynamic can be applied to predict possible reactions of metals in solution. Figure 4.4 is a pourbaix diagram which is shown a chemical reaction products of iron which is immersed in water. Form the graph, it can be detected chemical reactions involved when iron locates in water at various acidity conditions [1, 3].

When an element or substance changes from the first state (origin) to the second state (products), the free energy will change. There are three possible free energy changes during the reaction process [5, 1, 3].

- $\Delta G = Gf - Gi < 0$ (negative). The reaction process will take place spontaneously. Where the energy in the initial state (Gi) is greater than the energy in the final state (Gf).
- $\Delta G = Gf - Gi > 0$ (positive), the reaction occurs not spontaneously. The final energy (Gf) is higher than the initial energy (Gi).
- $\Delta G = Gf - Gi = 0$, the reaction reaches equilibrium, the energy does not change when a chemical process occurs.

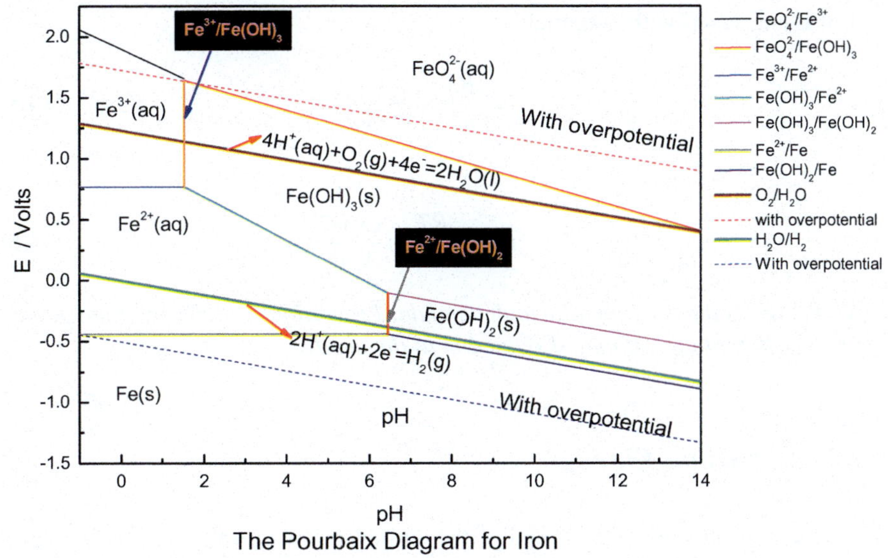

Fig. 4.4 Iron in water pourbaix diagram (*M. Pourbaix*). It shows the chemical reactions involved when iron locates in water at various acidity conditions [1], 3].

Under standard conditions: $T = 298\ k$, $R = 8.3143\ J\ (mol.k)^{-1}$, the standard free energy of the cell reaction is denoted as *Go* which is affected by the standard cell potential difference, *Eo*, expressed by the formula:

$$\Delta Go = -nF\Delta Eo \tag{4.1}$$

where E is the Sell Potential (Volts), F: Faraday Constant (coulomb/mol.E), N: Number of Electrons (mol).

If, in solution, a chemical reaction occurs between component *A* and component *B* and forms components *C* and *D* with respective concentrations *a, b, c, d*. Chemically it can be written as follows [5, 6]:

$$aA + bB \rightarrow cC + dD \tag{4.2}$$

Calculation of free energy that occurs is:

$$\Delta G = \Delta G^o + RTln\frac{C^c D^d}{A^a B^b} \tag{4.3}$$

If the reaction occurs at equilibrium, then $\Delta G = 0$, so it can be written as:

$$\Delta G^o = -RTln\frac{C^c D^d}{A^a B^b} \tag{4.4}$$

If it is expressed with potential:

$$E^o = \frac{RT}{nF} ln \frac{C^c D^d}{A^a B^b} \tag{4.5}$$

or

$$E = E^o - \frac{RT}{nF} ln \frac{C^c D^d}{A^a B^b} \tag{4.6}$$

With, ΔEo is the cell potential under standard conditions which can be calculated by: Ereduction at the cathode–Ereduction at the anode.

4.4.2 Corrosion Kinetics

Kinetic corrosion studies the speed of the corrosion reaction that occurs in metals during the corrosion process. The corrosion reaction that occurs on the surface between the metal-water (electrode–electrolyte) produces a current density of electrons with a certain speed. Figure 4.5 is a schematic of an electrochemical cell when a metal is immersed in an acid solution containing O_2 gas [2, 6, 7].

The speed of the flow of electrons is determined by the potential difference between the two electrodes. In electrochemistry, the writing of metal oxidation reactions is as follows:

$$M \rightarrow Mn^+ + ne^- \tag{4.7}$$

This equation means that if a metal dissolves in water, each metal atom will release 2 electrons. According to Avogadro, one mole of electrons contains 6 × 1023 atoms.

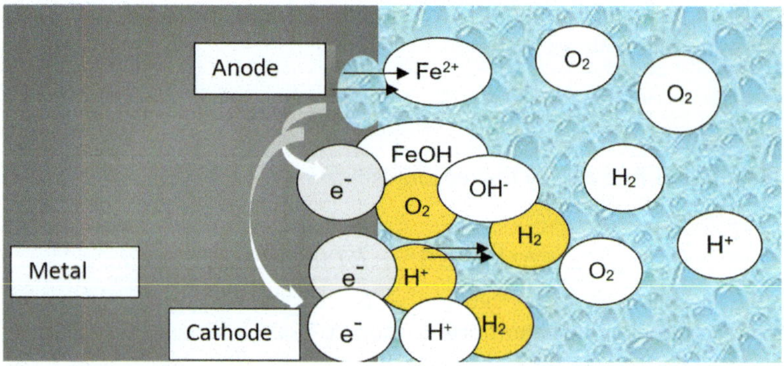

Fig. 4.5 Schematic of the formation of an electrochemical cell when a metal in an acid solution containing O_2 gas undergoes a corrosion reaction

And 1 mol of electrons has a charge of 96,500 °C (called the constant *Faraday* (*F*)). Thus the amount of electric charge flowing amounted to:

$$Q = \frac{nFm}{M} \tag{4.8}$$

where Q: electric charge, n: number of electrons moving, F: Farady constant, M: molecular weight.

Faraday's law states that the number of exchanges of electrons through the electrolyte per unit time is called the current (amperes). Thus we can rearrange Eq. (4.8) by dividing by time to:

$$I = \frac{nFW}{M} \tag{4.9}$$

where $I =$ current (Ampere), $W =$ weight loss (g/sec).

When both sides are divided by the exposed area of the body, the formula for the corrosion rate can be represented by the thickness reduction formula, namely:

$$CR = k\frac{Mi}{nF\rho} \tag{4.10}$$

where $CR =$ corrosion rate *(mm/sec)*, $i =$ current density *(Amperes/m2)*, $\rho =$ density *(g/m³)*, $k =$ constant value.

4.4.3 *Corrosion Rate*

Measuring the corrosion rate can use the electrochemical formula developed by the *Stern-Geary* equation. When a metal undergoes a corrosion process, the dominant factor affecting the corrosion rate is the anodic reaction where an oxidation process occurs. Thus, from the thermodynamic formula, the polarization behavior can be expressed by the corrosion potential [1]:

$$E = E^o + \frac{RT}{nF} ln\left(\frac{i_o}{i_o}\right) \tag{4.11}$$

Or:

$$i_a = i_o \exp^{\eta}/_{b_a} \tag{4.12}$$

where, $\eta =$ E–Eo and ba $=$ RT/αnF or the slope that occurs at the anode.

At the cathode site, the corrosion reaction is controlled by the reduction process. Electrochemically, the reduction reaction can be expressed as:

$$i_c = i_o \exp^{-\eta}/_{b_c} \tag{4.13}$$

The total current density is obtained by calculating the current difference between the anode and cathode currents.

$$i = ia - ic \tag{4.14}$$

Equations 4.12 and 4.13 are substituted into Eq. 4.14 to become:

$$\boldsymbol{i} = i_o\left[exp\left(\frac{\eta}{b_a}\right) - exp\left(\frac{-\eta}{b_c}\right)\right] \tag{4.15}$$

At a location around E = 0, $\eta/b = 0$ and exp(η/b) = $(1 + \eta/b)$, so that Eq. 4.15 becomes:

$$\boldsymbol{i} = i_o\left[\left(1 + \frac{\eta}{b_a}\right) - \left(1 + \frac{-\eta}{b_c}\right)\right] \tag{4.16}$$

Which can be simplified to:

$$i = i_{corr}\eta\left(\frac{\beta_a + \beta_c}{\beta_a\beta_c}\right) \tag{4.17}$$

Furthermore Eq. 4.17 can be used to determine the resistance due to polarization which is defined as:

$$\frac{\eta}{i} = \left(\frac{\beta_a\beta_c}{i_{corr}(\beta_a + \beta_c)}\right) = R_p \tag{4.18}$$

And it can be simplified to:

$$i_{corr} = \frac{B}{R_p} \tag{4.19}$$

With:

$$B = \left(\frac{\beta_a\beta_c}{\beta_a+\beta_c}\right) \tag{4.20}$$

Usually, B is assumed to be between *12 and 50* mV depending on the corrosion mechanism. If both processes are reactions activation B is $26\ mV$. if a mass transport reaction occurs B assume $52\ mV$. To determine the values of βa and βc, a larger polarization scan is performed (Tafel extrapolation) as shown in the polarization diagram [1, 3].

4.5 Electrode Potential

In electrochemical cell, there are two types of potential, namely the anode potential and the cathode potential. The anode potential is the potential value of a metal acting as the anode and the cathode potential is the metal potential at the cathode. At the anode an oxidation reaction occurs (releasing electrons) and at the cathode a reduction reaction occurs (accepting electrons), here is referred to as the oxidation potential and reduction potential. Potential between two metals can be measured using a potentiometer. These potential records the energy per unit charge used to create oxidation/reduction reactions. The potential of the cell indicates its ability to release electrons which is referred to as the oxidation potential which is located at the anode location. The same thing happened at the cathode. Furthermore, the cell potential value is expressed by the formula:

$$Ecell = oxidation\ potential + reduction\ potential$$

Or

$$\begin{aligned} Ecell &= reduction\ potential(at\ the\ cathode\ location) \\ &- reduction\ potential(at\ the\ anode\ location) \end{aligned}$$

Table 4.1 shows the value of the reduction potential at the cathode where if the potential value is more positive it means that the ability to experience reduction process is greater or the metal will tend to react reductionly. Conversely, the more negative the reduction potential value, the smaller the ability to undergo reduction or the metal will undergo oxidation. As an example of a standard measurement of the iron reduction potential is as follows. When iron is immersed in an acid solution, the reactions that occur consist of:

- Soluble iron:

$$Fe = Fe^{2+} + 2e^- \text{ (oxidation reaction at the anode)} \tag{4.21}$$

- Hydrogen gas produced from an acid solution:

$$2H^+ + 2e^- = H_2\text{(Reduction reaction at cathode)} \tag{4.22}$$

- overall reaction:

$$Fe + 2H^+ = Fe^{2+} + H_2 \tag{4.23}$$

Using the Nernst equation, it is obtained:

Table 4.1 The standard reduction cell potential was measured using a hydrogen reaction reference [3]

Electrodes (SHE)	Standard electrode potential, Eo V	
$Au^{3+} + 3e^{-} = Au$	1.5	
$Cl_2 + 2e^{-} = 2Cl^{-}$	1.36	
$O_2 + 2H^{+} + 2e = H_2O$	1,228	
$Br_2 + 2e = 2Br\_$	1,065	
$Ag^{+} + e = Ag$	0.799	More stable
$Hg_2{}^{2+} + 2e = 2Hg$	0.789	
$Fe^{2+} + e\_ = Fe^{3+}$	0.771	
$I_2 + 2e = 2I^{-}$	0.536	
$Cu^{+} + e = Cu$	0.52	
$Cu^{2+} + 2e = Cu$	0.337	
$2H^{+} + 2e = H_2$	**0.000 (standard reference)**	
$Pb^{2+} + 2e = Pb$	−0.126	
$Sn^{2+} + 2e = Sn$	−0.136	
$Ni^{2+} + 2e = Ni$	−0.250	
$Fe^{2+} + 2e = Fe$	−0.440	
$Cr^{3+} + 3e = Cr$	−0.740	More active
$Zn^{2+} + 2e = Zn$	−0.763	
$Al^{3+} + 3e = Al$	−1,663	
$Mg^{2+} + 2e = Mg$	−2,370	

$$E = E^{o} - \frac{0.059}{2} log \frac{Fe^{2+} H_2}{Fe(H^{+})^2} \tag{4.24}$$

If the values for H^{+} and H_2 and Fe^{2+} are 1 mol and Fe is a solid, the standard Fe potential can be calculated based on the hydrogen reaction. That is:

$$E = Eo = +0.44 \; V. \tag{4.25}$$

Based on the formula $G = -nFE$, the G reaction of iron in acid gas <0, so the reaction will occur spontaneously. In the same way, the standard electrode potential of other metals can be measured and the values obtained are arranged in Table 4.1.

4.6 Corrosion Risk Management

Corrosion is unavoidable. But corrosion can be calculated and predicted (*APIRP 581*). The purpose of managing corrosion can be classified into several actions, namely, identifying structural parts that are experiencing corrosion, determining the causes of damage, calculating when damage will occur and determining the choice of what method will be used to extend the life of the building structure and minimize costs [8]. To achieve that goal, measurable steps must be taken. The following is an example of steps to manage corrosion damage.

Step 1: Identify Risks

By defining the external and internal factors that cause damage it will be easy to manage risk. While at the same time establishing the scope and risk criteria for appropriate action to be taken. Identification of damage is done by reviewing the possibility and frequency of failure. Failures may include coating damage, corrosion, cracking, and failure of corrosion mitigation such as cathodic protection and corrosion inhibition. Consequence assessments address health and safety, environmental and financial/economic impacts.

Step2: Risk Ranking

*N*amely analyzing the type of threat, and the severity of the consequences. This assessment can be completed using a risk rating methodology. Risk is defined as the possibility of failure and the consequences of that failure. The risk assessment approach is used to assess risk through a semi-quantitative approach indicated by the Risk Assessment Matrix (RAM), as in Table 4.2 below. Level of risks are categorized as low, medium, and high which are notified with color coded or scored. Each Company can develop their own matrix that reflects their tolerance for risk in managing assets.

Step 3: Risk Evaluation

It is a process where the results of risk analysis are compared with the level of readiness to tackle the risks. From this comparison, it can be concluded how to address the level of risk that occurs.

Step 4: Risk Treatment

It is the application of a form of mitigation to overcome threats or accept the consequences of risk if the risk exceeds an acceptable level. The type and level of mitigation treatment must be related to the level of risk determined by calculating the impact of the costs incurred. As a follow-up to step 3, further steps are taken to reduce risk. This step ranges from replacement, repairing, maintenance, monitoring and inspection. Treatment depends on the type of asset in order to optimize and save costs.

Table 4.2 Risk assessment matrix analysis

			Impact Levels				
Likely to happen	score		1	2	3	4	5
			Can be ignored	Low	Intermediat e	Damaged	Badly damaged
	5	Often					
	4	Rather					
	3	Possible					
	2	Seldom					
	1	Very rarely					

Point	risk levels	Action
	Can be ignored	No action needed
	Low	Check and monitor
	Intermediate	Reduce work activity
	Tall	Fix it immediately
	Unacceptable	Stop process

Step 5: Risk Monitoring

It is the final stage of the risk management process and will be arranged for a reassessment of risk management, either with Key Performance Indicators (KPIs), or with indicators of achievement time.

References

1. Fontana, M.G.: Corrosion Engineering, 3rd edn. New York, McGraw-Hill (1986)
2. Salas, B.V., Wiener, M.S.: Corrosion Engineering: Principles and Practice. McGraw-Hill, USA (1991)
3. Uhlig, R.W.: Corrosion and Corrosion Control: An Introduction to Corrosion Science and Engineering, 3rd edn. Wiley, New York (1985)
4. Shreir, R.A.: Corrosion (two volumes), Jarman, and GT Burstein (eds.). Butterworth-Heinemann, Oxford, Boston (1994)
5. Davis, J.P.: Corrosion: Understanding the Basics, 2nd edn. Metals Park, USA, Ohio (2000)
6. Stansbury, E.E., Buchanan, R.A.: Fundamentals of Electrochemical Corrosion. ASM, Ohio, USA (2000)
7. Piron, D.L.: The Electrochemistry of Corrosion. NACE, Texas, USA (1991)
8. Morshed, A.: Corrosion Management and Associated Implementation Issues, MP 52, vol. 8 p. 66 (2013)
9. Morshed, A.: Corrosion-Management-and-Cost-Optimization, MP, 5, 4.APIRP 581. Risk-Based Inspection Technology (2020)

Chapter 5
Corrosion of Steel Reinforcement

Abstract Steel corrosion in concrete poses a substantial challenge within the realm of construction and infrastructure. This chapter focuses on the electro-chemical reactions of iron within concrete. It explores the fundamental processes that lead to steel corrosion. Understanding the factors causing corrosion of concrete reinforcement is crucial for developing effective preventive measures. The metallurgical properties of metals, including steel, play a vital role in their susceptibility to corrosion. Water conditions, such as moisture levels and the presence of aggressive substances, have a direct impact on the corrosion process. Specifically, the effect of chloride ions on reinforced concrete (RC) is a prominent factor that accelerates corrosion. In addition to chloride ions, the effects of CO_2, porosity, humidity, and pH levels within the concrete environment contribute to the corrosion of steel reinforcement. These factors can weaken the protective layer surrounding the steel, making it more vulnerable to corrosion. Understanding the complexities of steel corrosion in concrete and its underlying mechanisms is crucial for implementing effective corrosion prevention and mitigation strategies.

5.1 Corrosion of Steel in Concrete

Corrosion that occurs on the steel surface causes the addition of a larger volume on the steel surface, resulting in internal pressure on the concrete. If the corrosion continues, the concrete will be cracks, spall, and eventually, the structure damaged. Corrosion will occur if concrete is exposed to salt water. Corrosion of steel in concrete are often found in bridges/infrastructures built on rivers or seas. The corrosion process will be accelerated by the reaction of chloride with calcium carbonate and calcium aluminum. Carbonization reduces the pH value, and calcium carbonate compounds are formed in the form of calcite and aragonite in the concrete. In the seawater environment, corrosion is caused by chloride ions reacting with oxygen. The corrosion reaction in the steel will increase the volume of steel in the concrete which leads to create internal strees in the concrte as can be seen in Figs. 5.1 and 5.2. The corrosion processes are according to the reactions in Eqs. 5.1–5.6. The reaction of iron in concrete with water is formulated as follows [3, 11]:

Y. P. Asmara, *Concrete Reinforcement Degradation and Rehabilitation*,
Engineering Materials, https://doi.org/10.1007/978-981-99-5933-4_5

$$Fe^{2+} + 6H_2O \rightarrow \left[Fe(H_2O)_6\right]^{2+} \tag{5.1}$$

$$\left[Fe(H_2O)_6\right]^{2+} \rightarrow \left[Fe(OH)(H_2O)_5\right]^{+} + H^{+} \tag{5.2}$$

$$\left[Fe(OH)(H2O)_5\right]^{+} \rightarrow Fe(OH)_2 + 4H_2O + H^{+} \tag{5.3}$$

The presence of oxygen/chloride ions causes hydrated Fe^{2+} ions to form the iron oxide FeO-OH, according to the following reaction:

$$2Fe(OH)_2 + 1/2O_2 \rightarrow 2\gamma - FeOOH + H_2O \tag{5.4}$$

Another expression for the reaction of steel in concrete is given as follows [10]:

Fig. 5.1 Mechanism of damage to concrete due to the corrosion reaction that occurs in steel reinforcementvide

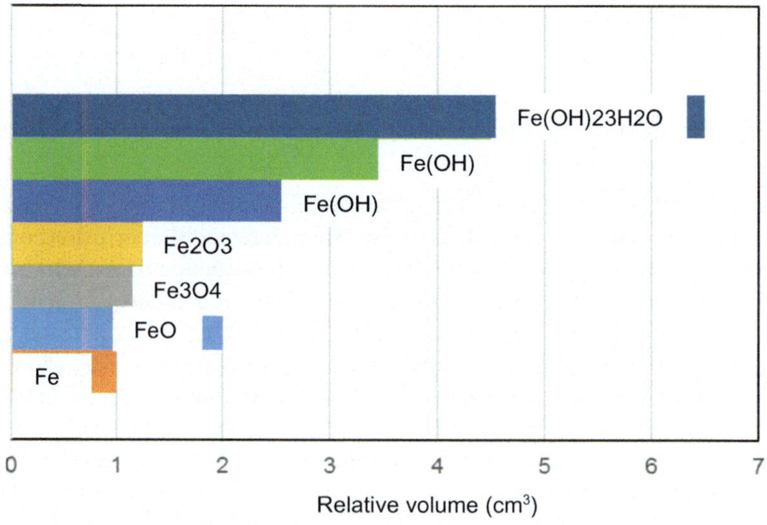

Fig. 5.2 Comparison of the volume of ferrous corrosion products inside concrete [8, 16]

$$xFe^{2+} + yO_2 + H_2O \rightarrow Fe_3O_4,\ Fe(OH_3),\ Fe(OH_3) \cdot 3H_2O \tag{5.5}$$

If O_2 is present, the overall reaction becomes:

$$2Fe + O_2 + 2H_2O \rightarrow 2(FeO(OH)) + H_2 \tag{5.6}$$

FeO(OH) is the mineral lepidocrocite. The expansion of Fe into the mineral lepidocrocite causes the volume to increase by 300%. In seawater, magnesium can replace the calcium in calcium hydroxide to form brucite ($Mg(OH)_2$), which will expand and form long crystals of aragonite or vaterite. Furthermore, chlorine ions will also change the chemical properties of concrete by forming calcium chloroaluminate ($Ca_3Al_2O_6CaCl_2 10H_2O$). Chloride ions will continue to penetrate to form crystals of sodium chloride (halite) and calcium chloride. The alkaline carbonate reaction involves a de-delomination reaction as in Eq. 5.7:

$$CaMg(CO_3)_2 + 2MOH \rightarrow Mg(OH)_2 + CaCO_3 + M_2CO_3 \tag{5.7}$$

where M stands for sodium, potassium, or both.

For steel that is immersed in water, the iron atom will lose its electrons to become positively charged ions, which dissolve into the electrolyte, and while acting as the cathode, a reduction reaction occurs on H^+ in aqueous solution, forming H_2 (see Fig. 5.3). The complete reaction is as follows [5, 14].

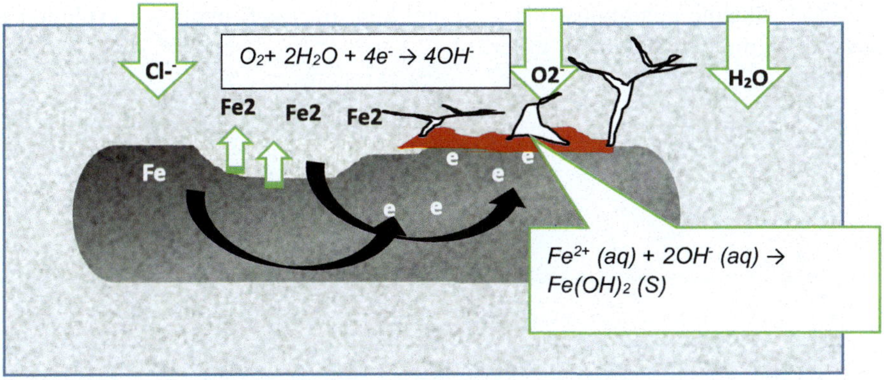

Fig. 5.3 Corrosion mechanism of steel reinforcement in concrete [4, 7]

5.1.1 *Electrochemical Reaction of Iron in Concrete*

- Reaction at the Anode

Iron dissolution occurs, where iron releases a number of electrons in an acid solution:

$$Fe \rightarrow Fe^{2+} + 2e^{-} \tag{5.8}$$

- *Reaction at the Cathode*

On the cathode side, there are several reactions that occur, depending on the state of the current. In an acidic environment, hydrogen evolution will occur, which can be explained as follows:

$$2H^{+} + 2e^{-} \rightarrow H_2 \tag{5.9}$$

Electrons can react with water and oxygen in the concrete on the concrete surface through the following reaction:

$$O_2(g) + 2H_2O\,(l) + 4e^{-} \rightarrow 4OH^{-}(aq) \tag{5.10}$$

For concrete contaminated with CO_2 (concrete carbonation), the reaction on the concrete is:

$$Fe + CO_2 + H_2O \rightarrow FeCO_3 + H_2 \tag{5.11}$$

In the presence of calcium ions, the combined reaction between CO_2 and calcium ions at the cathode becomes:

$$Ca(OH)_2 + CO_2 + H_2O \rightarrow CaCO_3 + 2H_2O \tag{5.12}$$

Under alkaline conditions (pH > 5) and high pressure, hydrogen evolution is possible by direct reduction of water:

$$2H_2O + 2e^{-} \rightarrow H_2 + 2OH^{-} \tag{5.13}$$

5.2 Factors Causing Corrosion of Concrete Reinforcement

Corrosion of steel in the concretes are influenced by several factos as presented in Table 5.1. Each factors determine rate of steel damages and pattern of damages. Table 5.2 shows comparison corrosion rate of carbon steel (CS) and stainless steel (SS316) at amine solution.

Table 5.1 Factors causing corrosion of reinforced concrete [9]

Steel reinforcement	Concrete quality
• Metallurgy • Surface quality • Steel size • Steel construction • Expansion coefficient	• Type of cement • Water/cement ratio • Water content • Aggregate type • Permeability • Concrete quality
Steel placement position	**Environmental conditions**
• The thickness of concrete cover • Steel distance • Concrete density • Compression • Concrete-steel adhesion • Drying	• Water vapor • Temperature • pH • Chloride • Oxygen • CO_2 • Cracks • External voltage • Loading

Table 5.2 Corrosion rate of carbon steel (CS) and stainless steel (SS316) at the location of the heat exchanger and amine [1]

Coupon exposure location	Alloys	Mass difference	Exposure time (h)	Area (cm^2)	Corrosion rate (mpy)	General observations
101 heat exchangers	CS	0.2343	2,009	12.93	3.95	General corrosion
	CS	0.2162	2,009	12.93	3.65	General corrosion
102 heat exchangers	CS	0.3469	2,009	16	4.37	General corrosion, black corroded film
	CS	0.012	646	10.73	0.676	Not corroded
	CS					
	SS316					
Amen	CS	0.2259	1,993	24.43	2.036	General corrosion, black corroded film
	SS316	0	1,993	18.08		Not corroded

5.2.1 Metallurgical Properties of Metals

The speed of metal corrosion in concrete is affected by the type of metal embedded in the concrete. The metal that is often used for reinforcement is carbon steel. There are many types of carbon steel. Each type of steel contains different elements to obtain the desired metallurgical properties. Elements combined with steel provide different corrosion resistance. The corrosion inhibitor element that is usually used is chromium (Cr). Chromium steel has good corrosion resistance, and the corrosion rate of Cr steel is very slow. The combination with other elements (e.g. nickel, copper, magnesium, and molybdenum) will further increase the corrosion resistance of steel. Corrosion that occurs in steel in concrete is caused by local metallurgical differences. The difference in potential energy encourages electrons to move from a high oxidation potential to an area of low oxidation potential. The energy contained in the metal is concentrated in the grain boundary area, which consists of impurities, phase shifts, and metal joints. Formation of corrosion cells are related to the atomic structure of steel, and the surface of the reinforcement. Other factors such as surface roughness, scratches, cuts, and slag also initiate corrosion process.

5.2.2 Effect of Water Conditions

In seawater, the rust deposits are dark red, and there is a deep black coating on the metal surface. This precipitate is caused by the presence of magnesium in the seawater. Meanwhile, in well water, the rust deposits are whitish-yellow. This whitish color is due to lime deposits in well water and chlorine deposits in fresh water. The corrosion rate for the steel submerged in water is affected by the interaction of various factors. In fresh water, the corrosion rate of 0.05 mm per year is common, although the rate may drop to 0.01 mm per year when limescale deposits have formed. In seawater, the average corrosion rate is in the range of 0.1–0.15 mm per year. The factors affecting the corrosion of carbon steel in seawater are presented in Table 5.3.

5.2.3 Effect of Chloride Ions on Reinforced Concrete

Chloride ion is a negatively charged ion (anion). The concentration of anions in electrolytes plays a very important role in affecting the properties of electrolytes. Chlorine ions attack the protective layer (i.e., the protective oxide film that forms by itself on the surface of the steel). Chloride ions with ferrous ions form ionic bonds. The greater the number of chloride ions in the solution, the more steel will decompose into ferrous ions to form ferric chloride compounds. The process of decomposing steel into iron ions is followed by the release of electro iron. Effect of ion chloride concentration on corrosion rates are presented at Figs. 5.4 and 5.5.

Table 5.3 Effect of seawater conditions on the corrosivity of carbon steel

Internal factor seawater	Influence on iron and steel
Chloride ions	Very corrosive to ferrous metals. Carbon steel and ordinary ferrous metals cannot be passivated (generally, the sodium chloride (NaCl) content in the sea is about 3.5%)
Electrical conductivity	The high conductivity causes electrons to flow from the anode to the cathode easily; hence, this will increase corrosion
Oxygen	Corrosion in steel is mostly cathodically controlled. Oxygen can depolarize the cathode, facilitating attack; thus, high oxygen content will increase corrosion
Speed	Corrosion rates increase, especially when there is alternating flow. Turbulent seawater results in (1) breaking down the rust barrier layer and (2) bringing in more oxygen. In addition, water impacts accelerate the penetration and cause cavities in the concrete, increasing the exposed steel surface
Temperature	Increasing temperature tends to accelerate corrosion attack. However, the influence of temperature can also inhibit the rate of corrosion. Hot seawater precipitates a protective layer of crust, vaporizing the attached oxygen; therefore, there will be less attacks
Dirt (biological)	Dirt of biological elements, where crustaceans tend to lower the attack because it blocks the entry of oxygen
Pollution	Sulfides contained in polluted seawater greatly accelerate the corrosion attack of steel

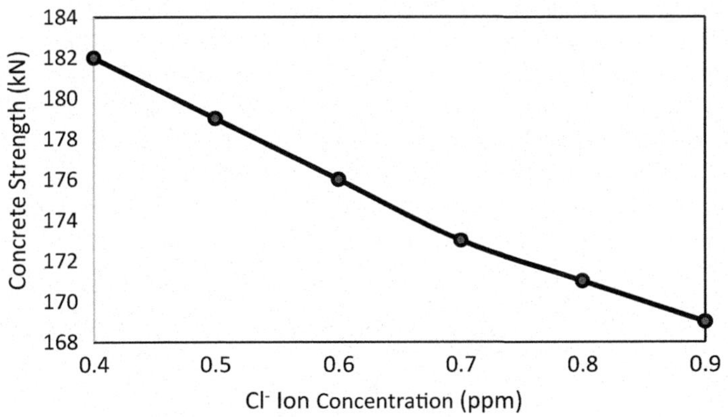

Fig. 5.4 Effect of the amount of chloride ions on the strength of RC [2]

The corrosion rate in reinforcing steel depends on mainly the Cl^-/OH^- ratio. The protective film can be destroyed by Cl^- even if the pH value is far above 11.5. When the Cl^-/OH^- molar ratio is higher than 0.6, the steel is no longer protected. If the iron oxide film is permeable or unstable, more oxygen can contact the iron surface. The presence of dissolved oxygen in concrete results in more oxygen being reduced

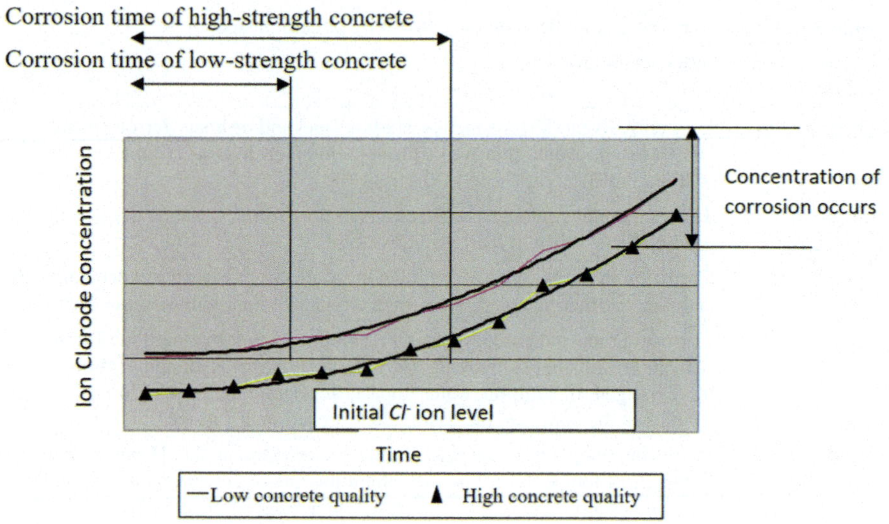

Fig. 5.5 Effect of concrete quality on the time required for corrosion to occur due to chloride ions [13]

to hydroxyl ions. This reaction consumes electrons generated from the oxidation in steel.

Chlorine ions damage concrete because they can increase the conductivity of concrete. In conductive concrete, the corrosion will be faster due to the high exchange current density. The damaging effect of chlorine ions occurs when its concentration reaches a critical value, which is about 0.4% by weight of the cement contained in the concrete. The corrosive nature of the chlorine ion is indicated by the diffusion coefficient. Diffusion determines the flow of passing electrons as an effect of the different concentrations in the medium. When the concentration of ions on the surface is higher, there will be a mass flux of chlorine ions into the concrete. The flux rate depends on concrete properties, such as thickness, integrity, and porosity.

$$Fe^{2+} + Cl^- \rightarrow FeCl_2 \quad (5.14)$$

$$FeCl_2 + 2H_2O \rightarrow Fe(OH)_2 + 2Cl^- \quad (5.15)$$

$$6FeCl_2 + O_2 + 6H_2O \rightarrow 2Fe_3O_4 + 12H^+ + 12Cl^- \quad (5.16)$$

The presence of Cl ions can change the mechanism of anodic iron dissolution. Along with changes in concrete conductivity, the rate of reaction of Cl ions with reinforced concrete (RC) is determined by the value of the diffusion coefficient. In plain concrete, migration will be faster when the concrete is porous. Chloride ions in RC cause pitting corrosion in RC. Chlorine ions affect the disintegration of concrete as an effect of crack propagation in concrete. The role of chloride in accelerating the corrosion rate is formulated in Eqs. 5.14–5.16. As seen from the equation, chlorine ions react with the steel repeatedly and continuously corrode the RC even in the presence of only small amounts of chloride. Free chloride can also act as a reaction catalyst that makes the corrosion process lasts for a long time. Chlorine ions can cause chemical changes in Portland cement, which cause a loss of strength [2].

5.2.4 *Effect of CO_2*

Carbon dioxide dissolved in seawater can react with water to form carbonic acid (H_2CO_3). Carbonic acid dissociates into hydrogen ions (H^+) and bicarbonate ions (HCO_3^-). The pH of the concrete is alkaline, but because seawater absorbs more CO_2, the pH of the concrete will decrease to become acidic. Figure 5.7 shows the dependence of reinforcement strength on CO_2 concentration [2]. A 50% mol% increase in the CO_2 content of the RC resulted in a reduction of 160 MPa over five years of service life. In further investigation, the decrease in strength is attributed to the RC depassivation as an effect of the acid reaction. In a CO_2 environment, the pH is affected by dissolved ferrous bicarbonate. The pH will decrease if there is an increase in bicarbonate ions. The simultaneous reaction of lowering the pH and increasing the corrosion rate can be explained by the properties of the protective film. At higher pH, concrete is passive due to scale formation on the RC. The higher the pH, the thicker, denser, and more protective the passive layer becomes. When the pH decreases, the rate of the anodic reaction increases. This observation is in accordance with the findings of [15] regarding the iron dissolution mechanism (see Fig. 5.6).

$$Ca(OH)_2 + CO_2 \rightarrow CaCO_3 + H_2O \tag{5.17}$$

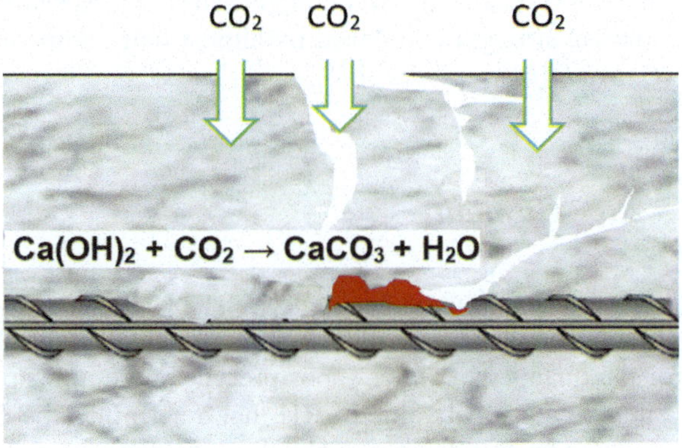

Fig. 5.6 Model of the corrosion mechanism of reinforcing steel due to CO_2 in RC [9]

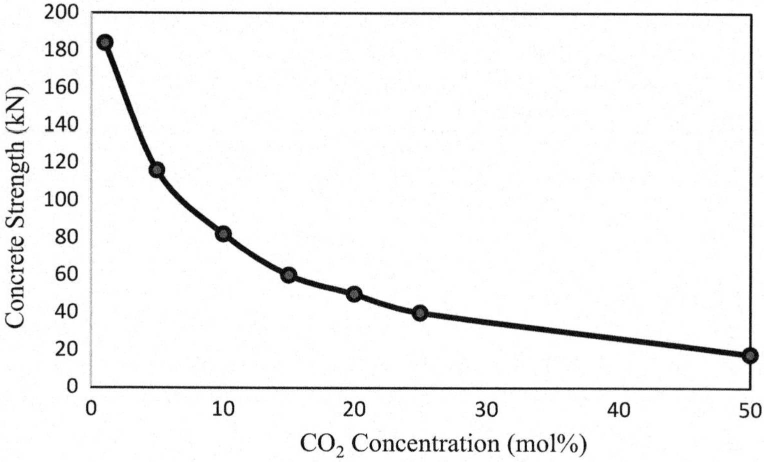

Fig. 5.7 Effect of the amount of CO_2 on the strength of RC [2]

5.2.5 Porosity Effect

Porosity causes the entry of corrosive elements to reach RC, which can reduce the corrosion potential and increase the rate of anodic and cathodic reactions. The current density of 0.08 A/cm^2 occurred at 30% concrete porosity [2]. The corrosion potential was reduced from −0.1 to −0.04 mV due to an increase in concrete porosity from 15 to 30%. The relationship between concrete porosity and the level of corrosivity of RC is related to the open access of water and oxygen to reach RC, which leads to oxygen reduction and hydrogen evolution [17]. Figure 5.8 shows the relationship between the porosity of concrete and the strength of concrete at various humidity levels (relative humidity (RH)). Increasing the porosity of the concrete decreases its strength. Figure 5.8 also presents the effect of RH on concrete strength, where higher RH results in lower concrete strength. At saturated RH and 30% porosity, the concrete strength is reduced by almost half compared to RH 81 with low porosity (*Gabriel* [6]).

5.2.6 Humidity

Figure 5.8 explains that RH can be associated with NaCl and other impurities that can dissolve on the surface of the reinforcement. Relative humidity is defined as the ratio of the quantity of water vapor present in the atmosphere to the saturated quantity. There is a critical value of RH that determines the corrosion process. The critical moisture level is determined by the corroded surface of the material. Higher porosity leads to a tendency for corrosion products and surface deposits to absorb moisture

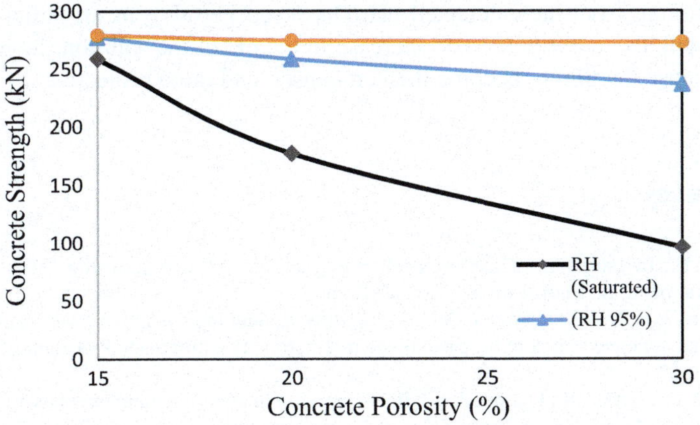

Fig. 5.8 Effect of porosity and moisture (RH) on RC strength (Romance, 2018)

Fig. 5.9 Trend corrosion of steel in aqueous solution as a function of OH^- ion concentration and pH level, with the reduced corrosion at pH 12.5 is a result of the formation of an effective passive film [12]

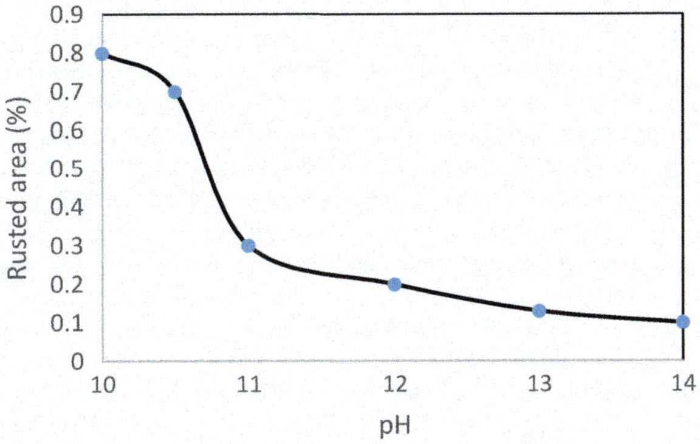

and a higher presence of atmospheric pollutants. In the presence of an electrolyte in the form of a thin-film layer, atmospheric corrosion takes place with balanced anodic and cathodic reactions. Anodic oxidation occurs on the surface of the metal, while the cathodic reaction is oxygen reduction (*Szweda Zofiaa, 2013*).

5.2.7 pH

pH is an important parameter in corrosion. The corrosion rate will be lower at higher pH (Fig. 5.9). The calculation of pH can be done by considering the kinetics of iron hydroxide ($Fe(OH)_2$) precipitation. The pH is also affected by the concentration of H^+ ions, temperature, pressure, and dissolved ions, which results in the actual pH being different from the calculated pH. Dissolved $Fe(OH)_2$ as an initial corrosion product will also contribute to an increase in the pH of the solution. Increasing the pH will cause the film to become thicker, denser, and more protective.

References

1. Amir, E.: Investigation of carbon steel and stainless steel corrosion in a MEA based CO_2 removal plant. In: Petroleum and Coal (2015)
2. Asmara, Y.P., Nor, M.I., Anwar, S.N.R. Sugiman: Remaining strength prediction of reinforced steel bar concrete structure in seawater environment. AIP Conference Proceedings, Vol. 2489, p. 1 (2022)
3. Bentur, A., Berke, N., Diamond, S.: Steel corrosion in concrete: fundamentals and civil engineering practice. Taylor & Francis Group. Created from intiuc-ebooks on 2022-12-06 00:08:37 (2011)
4. Broomfield, John, P.: Corrosion of Steel in Concrete: Understanding, Investigation and Repair, St. Edmundsbury Press Limited, Bury St. Edmunds, Suffolk, Great Britain, 238 (1997)
5. Fontana, M.G.: Corrosion Engineering, 3rd edn. McGraw-Hill, New York (1986)
6. Gabriel Samson: Fabrice Deby a, Jean-Luc Garcia, Mansour Lassoued, An alternative method to measure corrosion rate of reinforced concrete structures. Cement Concrete Compos. **112** (2020). https://www.researchgate.net/publication/320172742_Studies_on_the_Performance_of_Migratory_Corrosion_Inhibitors_in_the_Corrosion_Control_of_Concrete_Rebars
7. Ibrahim, M.: Studies on the Performance of Migratory Corrosion Inhibitors in the Corrosion Control of Concrete Rebar, October 2017, Materials Performance and Welding Technology Performance Conference, C-023 (2017)
8. Jones, D.A.: Principles and Prevention of Corrosion. Macmillan Publishing Company, New York, NY (1992)
9. PCA: Concrete Slab Surface Defects: Causes, Prevention, Repair, IS177, Portland Cement Association (2001)
10. Rosenberg, A.C., Hansson, Andrade, C.: Mechanisms of corrosion of steel in concrete. In: Berke, N., Chaker, V., Whiting, D. (eds.), Corrosion Rates of Steel in Concrete (West Conshohocken, PA: ASTM International), pp. 174–188 (1990). https://www.scipedia.com/public/Rosenberg_et_al_1990a
11. Sastri: Green Corrosion Inhibitors: Theories and Practice. Wiley (2011)
12. Shalon, R., Raphael, M.: Influence of sea water on corrosion of reinforcement. Am. Concrete Inst. **30**(12), 1251–1268 (1959)
13. Ueli, M.: Challenges and opportunities in corrosion of steel in concrete. Mater. Struct. **51**, Article number: 4 (2018)
14. Uhlig, Revie, R.W.: Corrosion and Corrosion Control: An Introduction to Corrosion Science and Engineering (3rd edition), Wiley, New York (1985)
15. Videm, K.: The effects of some environmental variabales on the aqueous CO_2 corrosion of carbon steel. Inst. Mater. **13** (1984)
16. Zemajtis, J.: Modeling the time to corrosion initiation for concretes with mineral admixtures and/or corrosion inhibitors in chloride-laden environments. Ph.D. diss. Va. Tech. (1998)
17. Zofia, Z.A.: An experimental study of the effects of chloride ions on the corrosion performance of polymer coated rebar in concrete pavement, May. J. Asian Archit. Build. Eng. **11**(1) (2012)

Chapter 6
Reinforced Concrete Corrosion Experiments

Abstract This chapter explores the techniques employed in corrosion testing, including accelerated corrosion experiments to simulate real corrosion process. Environmental simulation, such as immersion in simulated concrete pore solution, enables researchers to replicate the concrete's corrosive conditions. Electrochemical experiments, using tools like potentiostats, electrodes, and electric potential measurement, offer insights into the electrochemical processes involved in corrosion. Surface observation and visual inspection aid in identifying visible signs of corrosion damage. The analysis of corrosion products and dissolved elements using techniques like optic microscopy provides valuable information about the corrosion mechanisms and degradation products. Quantifying corrosion rate is crucial, which can be achieved through methods such as linear polarization resistance (LPR) and Tafel extrapolation. Electrochemical impedance spectroscopy (EIS) offers a comprehensive view of the electrochemical behavior of the system. To assess the extent of corrosion, techniques like corrosion potential mapping and attachment strength testing are employed. Penetration testing of chemical elements in concrete helps evaluate the impact of different elements on corrosion susceptibility. Finally, potential data reading provides critical information about the corrosion behavior over time. This knowledge contributes to the development of effective corrosion prevention and mitigation strategies, ensuring the durability and longevity of concrete structures.

6.1 Types of Corrosion Experiments

Corrosion occurs continuously in every construction, product, and facility made of metal. The metal decays faster as it interacts with more chemical elements. A way to determine when a material will break down is to perform an experiment. Experiments are carried out to monitor the process of damage and take appropriate actions so that damage does not occur suddenly. Damage needs to be monitored so that industrial assets, facilities, and equipment can work according to plan. In order to run the production process smoothly, the speed of damage needs to be calculated and the cause of the damage must be known. Monitoring corrosion is one way to ensure that production meets deadlines.

Y. P. Asmara, *Concrete Reinforcement Degradation and Rehabilitation*,
Engineering Materials, https://doi.org/10.1007/978-981-99-5933-4_6

6.1.1 *Corrosion Testing Techniques*

There are four types of tests that can be performed to test the corrosion behavior of a material, namely destructive and non-destructive tests, each of which consists of direct (in the field) and indirect (in the laboratory or simulation) tests (see Fig. 6.1). Selection of the type of test is based on ease of implementation, ease of accessing test results, cost, length of testing, the complexity of the tools used, data interception capabilities, and ease of validation with actual results [9, 15].

Corrosion testing can also be divided based on the environment, namely testing using electrolytes and without using electrolytes. Corrosion data can be obtained electrochemically or by looking directly at the pattern of corrosion that occurs (Fig. 6.2).

Tests in the laboratory are carried out on workpieces in certain environments to simulate the actual conditions that occur in these objects. Measurable preparation is required before testing. In testing, the parameters to consider are:

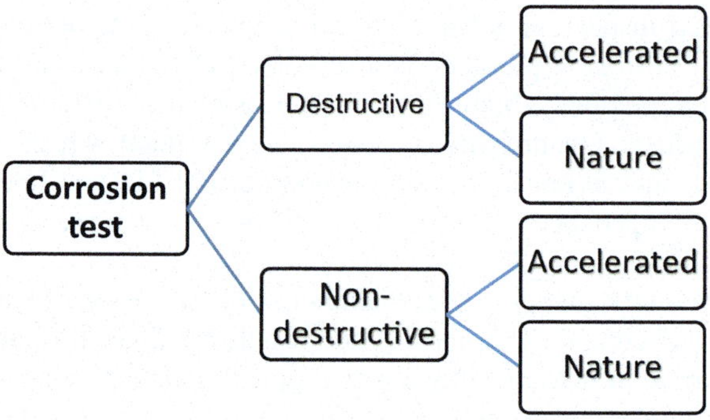

Fig. 6.1 Classification of corrosion testing

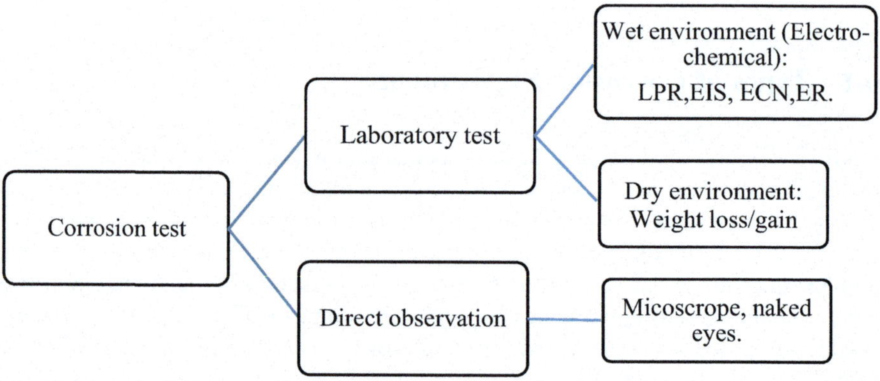

Fig. 6.2 Test methods to obtain corrosion velocity data

Environment:

- Dry (high temperature).
- Wet (immersion).
- Chemical composition.
- Mechanical loads (static and dynamic).

Test method:

- Electrochemical.
- Chemical analysis.
- Heat transfer.
- Immersion.
- Simulation (empirical, mechanical, and mathematical).

Result reading:

- Simulation models.
- Corrosion speed (addition, weight reduction, and thickness thinning).
- Thermal distribution.
- Dissolved chemical elements.
- Residual stress distribution.
- Electrical voltage distribution.

6.1.2 Accelerated Corrosion Experiments

The most ideal thing in an accelerated corrosion test is if the results obtained can give the same results as the tests carried out directly in the field. Thus, the parameters and conditions of the accelerated corrosion testing need to be as close as possible to the conditions that occur in the field. Factors that make it difficult to achieve the accelerated corrosion laboratory test results with the field test results are:

- The accelerated test is carried out by shortening the corrosion process time.
- Laboratory tests are simpler and standardized, with fewer variables than actual field conditions.
- Simulation of the failure mechanism of materials due to their interactions with the environment is difficult to simplify.
- Workpiece size, surface conditions, metal structure, sample geometry, and limited number of samples can affect laboratory studies compared to actual conditions.

Some examples of treatments that have been carried out to accelerate corrosion damage are:

- Changing the properties of environment thermochemistry increases the kinetics of corrosion.
- Speeding up the thermal cycle.

- Lowering the pH of corrosive media.
- Increasing the concentration of dissolved elements.
- Workpiece modification: artificial cracking, corrosion-prone geometry, and etching.
- Providing artificial electric current and voltage (electrochemically polarized test).
- Combining mechanical degradation (e.g., erosion, cavitation, and impingement) with corrosion damage.

However, applying the corrosion acceleration factors above will face the risk of obtaining results that deviate from the reality in the field. Accelerated corrosion tests in the laboratory have limitations as a result of the influence of the number of samples, sample geometry, chemical conditions, and mechanical problems leading to failure mechanisms. In other words, the test method does not represent complex variables under actual conditions.

6.1.3 Workpiece Simulation

The accelerated test workpiece may be a coupon, an electrode, a piece of pipe, or a sheet plate. The corrosion rate of the coupon is calculated using the weight loss or weight gain method. Meanwhile, for samples in the form of electrodes, the corrosion rate can be calculated by measuring the electrical properties of the workpiece, such as electric resistance (ER) which measures changes in resistance before and after the experiment. Linear polarization resistance (LPR), electrochemical noise (ECN), and electrochemical impedance spectroscopy (EIS) record the potential versus current at corroded electrodes to determine the resistance value due to corrosion, which can then be converted to corrosion velocity [13].

6.1.4 Coupon

Coupons are small samples made from actual materials in the field and can be checked periodically. Tests using coupons can be carried out in an accelerated or non-accelerated approach (test under actual conditions). The coupon is placed in an environment where the corrosion rate will be calculated, which can be read at any time by changes in surface shape and dimensions over a certain period of time. Coupons are designed according to the experimental purpose, for example, to test crevice corrosion and galvanic corrosion. Metal weight loss is an average representing uniform corrosion. The advantages of using coupons are simple, easy, cheap, and any environment can be easily simulated. However, the use of coupons has the disadvantages of the sample that can be damaged, long testing time, the activity that depends on the surface/volume, and relatively insensitive to local corrosion.

6.2 Environmental Simulation

6.2.1 Immersion

Immersion testing is a type of corrosion test that involves a corrosive solution to test corrosion rate of metals for a specified period. In an immersion test, a sample of the material is usually placed in a container filled with a corrosive solution, such as saltwater or acidic solution, and left for a certain period [3]. During this time, the material is exposed to the corrosive environment, and the corrosion rate is measured using various techniques. Figure 6.3 shows schematic immersion test to evaluate corrosion performance of materials.

Among the considerations in performing the immersion test include:

- A relatively simple test involving immersion at constant conditions. Acceleration of corrosion damage can be achieved from the length of immersion in corrosive media and other acceleration factors can be adjusted easily.
- The test can be carried out on a cycle test involving simultaneous soaking and drying.
- Instrumentation of the test specimen during immersion (e.g., connection to electrochemical instrumentation) must be considered to facilitate the measurement of changes in resistance.
- Testing can use a simple beaker.

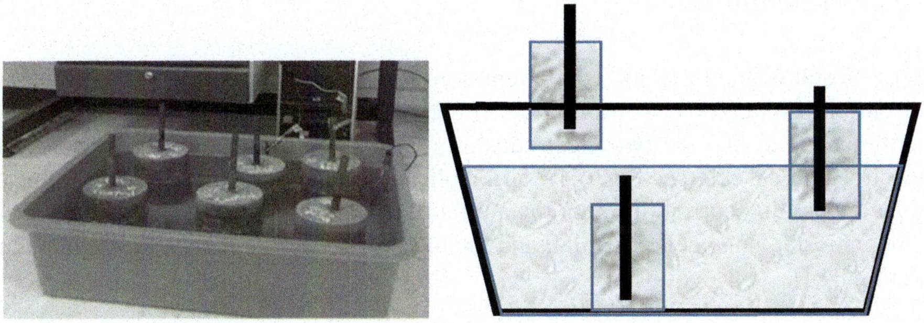

(a) (b)

Fig. 6.3 Concrete immersion test (**a**). The workpiece is placed on top, partially or completely submerged in the media under study and (**b**). the test is used to simulate the tide cycle where periodic wet and dry effects occur [2]

Table 6.1 The chemical composition of the solution commonly used to simulate concrete [10]

Chemical elements	Mole/L
NaOH	0.1
KOH	0.3
$Ca(OH)_2$	0.03
$CaSO_4-HO$	0.002

6.2.2 *Concrete Simulation*

Artificial concrete (simulated concrete pore solution) is made by making a solution containing chemical elements as shown in Table 6.1. The artificial electrolyte is assumed to resemble the behavior of concrete in the observed environment. The purpose of making an electrolyte solution is to make it easier to carry out research, such as taking aggressive ion migration samples, recording the dissolution of steel reinforcement in concrete, and measuring the distribution of chemical elements in concrete due to electric or magnetic fields. Concrete simulation can also use a saturated sodium hydroxide (NaOH) + $Ca(OH)_2$ solution.

6.3 Electrochemical Experiments

6.3.1 *Potentiostat*

Potentiostat is an electronic equipment used to measure the difference in potential or current between the workpiece (working electrode) and the reference electrode while changing it. The two electrodes are placed in a liquid medium to form an electrochemical cell (see Fig. 6.4). The application of a potentiostat to control this is done by passing current through the auxiliary electrode. The controlled variable in a potentiostat is the cell potential, and the measured variable is the electrochemical current [14].

6.3.2 *Electrodes*

Basically, there are three electrodes used in a potentiostat, namely the test electrode (working electrode (WE)), the reference/comparison electrode (RE), and the auxiliary electrode (AE or counter electrode (CE)).

- ***Working electrode***

In corrosion testing, the WE is the sample or material being tested. Usually, the metal WE is made of the actual material used in the field. This test is similar to the coupon

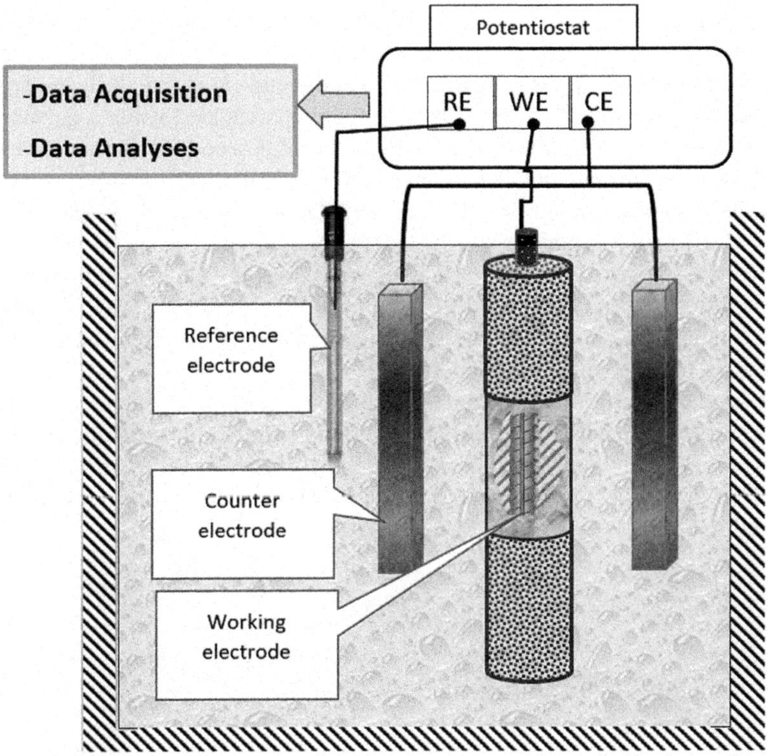

Fig. 6.4 Electrochemical corrosion of steel in concrete. The workpiece is steel in concrete, connected to the CE (auxiliary electrode) and the RE. All electrodes are connected to a potentiostat to collect electrochemical corrosion data

test. The WE can be pure or painted metal. Most of the electrochemical reactions that occur can be studied using this WE.

- ***Comparison Electrode (Reference Electrode)***

This electrode is used to measure the potential. The RE should have a constant electrochemical potential as long as the current is flowing. The electrodes commonly used are the saturated calomel electrode (SCE) and the silver chloride electrode (Ag/AgCl). Various types of electrodes are adapted to field conditions.

6.3.3 Electric Potential Measurement

If there is contact between the anode and the cathode, current will flow in the electrochemical cell. As shown in Fig. 6.5, corrosion can occur at the anode if there is a difference in free energy between the anode and the cathode. This energy difference is defined as the electric potential, which can be measured using a voltmeter. This potential can be interpreted as a tendency for corrosion to occur. If the circuit between the electrodes is closed, the potential to drive the current is the electrons produced by the corrosion reaction. Thus, corrosion can be calculated using a galvanometer that functions to measure current flow in a wet corrosion cell. All corrosion reactions in an aqueous environment can be considered those of a wet corrosion cell [12]. In the figure, it can be explained that if Zn and Cu are connected by cables, Zn will release electrons or oxidation will occur (the anode potential of Zn is higher than Cu). Electrons released through the cable are received by Cu, which makes Cu ions in the solution stick to Cu metal to become solid, while solid Zn will dissolve into Zn ions in the solution. The flow of electrons in a cable is an electric current that can be recorded using an amperage/voltmeter.

The level of damage to reinforcing steel can be determined by connecting the steel in the concrete with a RE (Fig. 6.6). The potential reading indicates the condition of the concrete, which can be related to the degree of deterioration of the reinforcing steel. The relationship between the damage to steel and the measured potential can be found in Chap. 3: Damage to Reinforced Concrete.

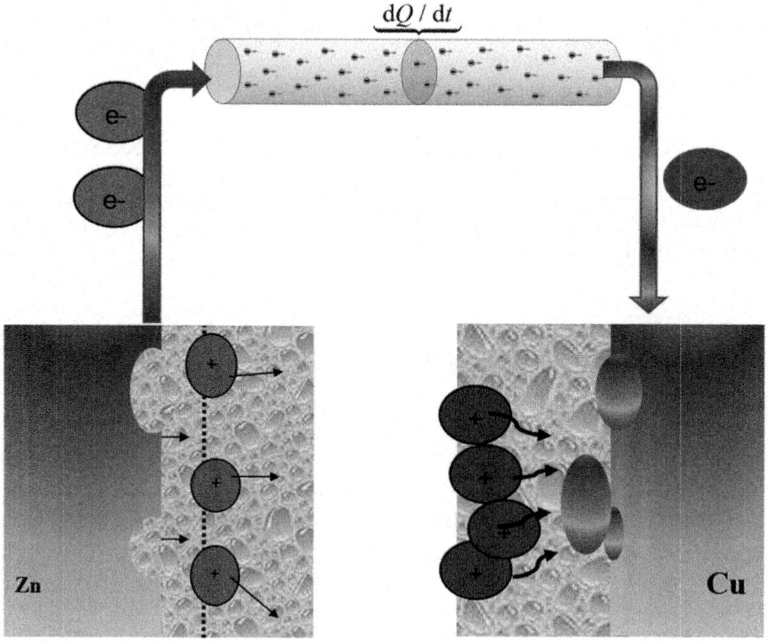

Fig. 6.5 The corrosion velocity measurement scheme uses an electrochemical cell that produces an electric current, and the amount of electricity that occurs can be obtained. Zn is the anode, where oxidation occurs and releases electrons to be given to Cu

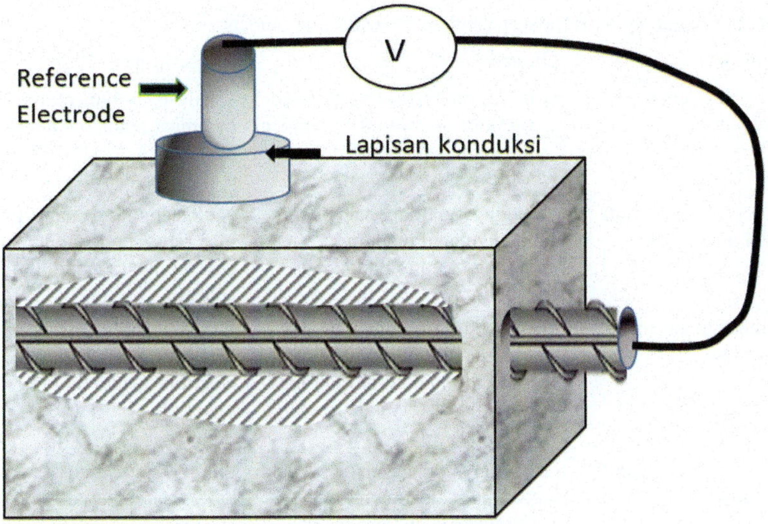

Fig. 6.6 Measurement of the potential of reinforcing steel in concrete. Steel is connected to a RE to record its potential value relative to the RE used. Based on the potential value, concrete is classified as corrosivity to iron

6.4 Surface Observation

Surface observation is the process of examining the surface of a material or object to identify any physical features, defects, or changes. Surface observation is typically performed using various tools and techniques, such as visual inspection, microscopy, or surface analysis [6]. Table 6.2 is presented equipment used for surface characterization of objects.

6.4.1 Visual Inspection

This method directly determines the occurrence of corrosion. This is a fundamental method for diagnosing or checking the shape and distribution of corrosion. Signs of corrosion damage can be immediately recognized, including corrosion due to cracks, leaks, discoloration, coating thickening, and bending construction, which can be implemented in analyzing the type of damage. Figure 6.7 is examples of images of Fe-1.3% C that shows microstructure of the sample after immersion test at certain period of time.

Table 6.2 Equipment used for surface characterization of objects

Equipment	Function	Application
X-Ray photoelectron spectrometer	Surface analysis of various solid materials, determining chemical state composition and depth profile	Catalysts, polymers, metals, ceramics, semiconductors, and electronics
Field emission scanning electron microscopy	Imaging, qualitative analysis, and elemental mapping; imaging of non-conductive samples	Nanotechnology, nanostructured materials, semiconductor development, and failure analysis
Transmission electron microscopy	High-resolution imaging technique that yields information about morphology, crystal phase, crystal structure, and defects	Catalysts, polymers, metals, ceramics, semiconductors, electronics, and biological materials
Universal scanning probe microscopy	Imaging and measuring surfaces on a fine scale, down to the level of molecules and groups of atoms	Polymers, ceramics, catalysts, biological materials, and electronics

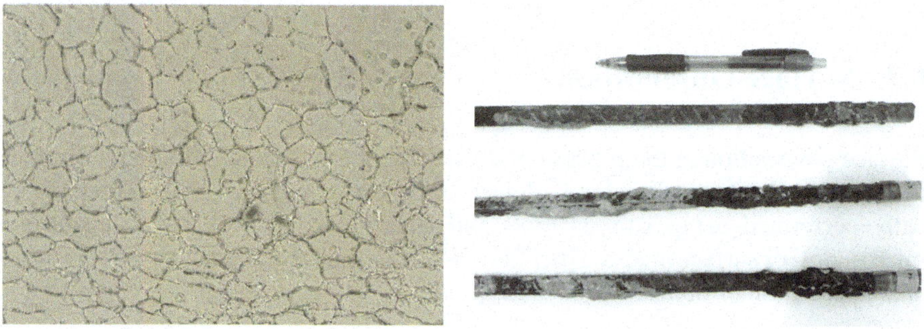

(a) (b)

Fig. 6.7 The image of Fe-1.3% C taken with an optical microscope (**a**) and direct observation of the corrosion of the steel reinforcement in the concrete after the immersion test (**b**)

6.4.2 Analysis of Corrosion Products and Dissolved Elements

The method is carried out by testing the chemical composition of the solution that is directly in contact with the observed material. It is assumed that during corrosion, the metal dissolves into the solution.

6.4.3 *Optical Microscopes*

Optical microscopes use one or a series of lenses to magnify an image using the light that is reflected from an object. Optical microscopes can be used to study metallography and the structure of metals and their alloys. Furthermore, to improve the resolution and increase the magnification, digital microscopes and stereo microscopes have been developed.

6.5 Calculating the Corrosion Rate

6.5.1 *Linear Polarization*

Corrosion is an electrochemical reaction as a result of the exchange of electrons between metals in solution. It is known that there are two equilibrium chemical reactions that occur in corrosion, namely the anode reaction and the cathode reaction. At the anode, oxidation occurs as the metal releases its electrons to be given to the cathode, which undergoes reduction. The reduction process involves the reduction of hydrogen, water, or oxygen ions. Corrosion rate is taken from the graph that relates potential and current density (Fig. 6.8).

A method for recording electrochemical events is by using a potentiostat. Potentiostats can record the relationship between the current and the potential that occurs on the metal surface. The description of the relationship between potential and current forms an equation curve called the Tafel equation curve, which is developed from the Butler-Volmer equation (Eq. 6.2). Electrochemical corrosion monitoring is based on

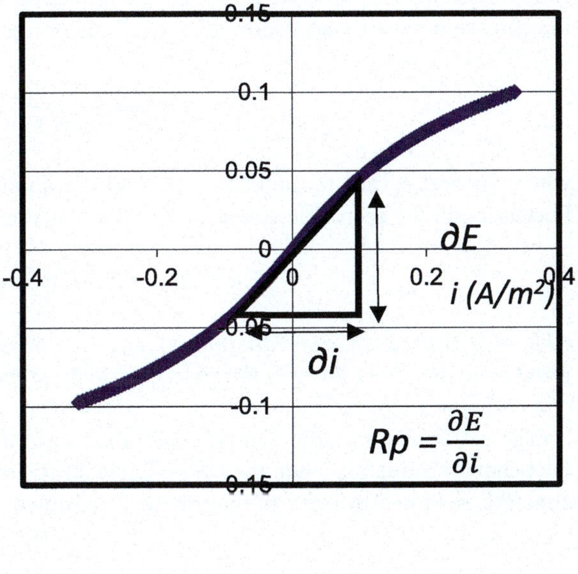

Fig. 6.8 The LPR diagram obtained by scanning the potential (changing the workpiece potential by $x \pm$ 10 mV from the equilibrium potential) [9, 14]

the premise that corrosion is basically an electrochemical process related to potential and current [9]. The corrosion speed is calculated using Eq. (6.1) as follow.

$$R_p = \frac{B}{i_{corr}} = \frac{(\Delta E)}{(\Delta i)_{\Delta E \to 0}} \tag{6.1}$$

where, R_p is calculated based on the formula obtained from scanning the difference in potential change divided by the change in current at the equilibrium point (E_{corr}).

Among the things to note in reading the corrosion velocity data with the LPR technique are:

- The form of corrosion is assumed to be uniform corrosion.
- The reactions that occur are anodic and cathodic reactions.
- Single anodic and cathodic reactions occur.
- The value of the Tafel constant must be known.
- Negligible solution resistance (i.e., solution with relatively high conductivity).
- Corrosion potential is calculated from a stable potential (equilibrium).

6.5.2 Extrapolation Tables

The corrosion behavior of a material based on the extrapolation Tafel method is tested by changing the corrosion potential in a positive or negative direction (polarization scan) to the corrosion potential. When the potential of an object changes, it will be followed by a change in current. The equation obtained is in accordance with the Stern-Geary equation. If the potential of the object is made more positive to the corrosion potential, it is said to be a change to the anode potential. Furthermore, the change toward the negative potential is called the cathode potential. The slope of the cathode and anode potential versus the current (slope) is used to obtain the value of the corrosion velocity according to the following Eq. (6.2).

$$I_{\text{corr}} = \frac{b_a b_c}{R_p 2.303(b_a b_c)}, \tag{6.2}$$

where b_a and b_c are the anode slope and the anode/cathode slope, respectively. Furthermore, the corrosion speed can be calculated using Eq. (6.3):

$$CR = I_{corr} 3272\, EW/\rho A \tag{6.3}$$

where CR is the corrosion rate (mm/y), I_{corr} is corrosion current (amps), EW is the gram equivalent weight, ρ is the specific gravity of metal (g/cm^3), and A is the area of metal (cm^2).

The extrapolation table can be used to measure the rate of corrosion and study the behavior of metals when they experience corrosion. The Tafel plot of reinforced concrete is shown in Figs. 6.9 and 6.10. It is noted that at polarization, there is a

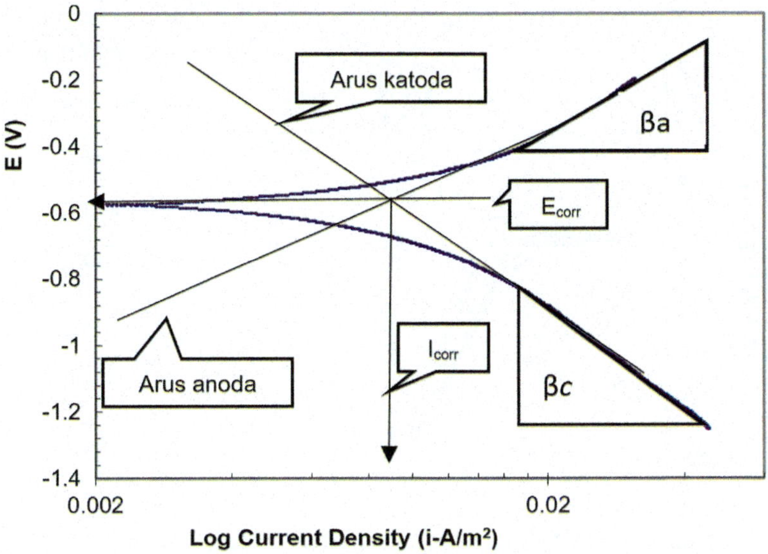

Fig. 6.9 Polarization scan diagram to calculate corrosion velocity using the Tafel extrapolation method [9]

decrease in corrosion current, where the reinforcing steel has been passivated at different levels. Concrete A experiences significant passivation compared to other concretes.

However, there are some errors that may occur when reading the Tafel curve, namely non-kinetic corrosion mechanisms, such as:

- The occurrence of concentration polarization, where the rate of reaction is affected by changes in the surface of the metal. At more negative polarization, the cathodic reaction will experience concentration polarization. This is due to the diffusion of oxygen or hydrogen formed by the reaction.
- If the polarization in a more positive direction results in a change in the metal surface, the corrosion decreases (becomes passive).
- Other effects that change the surface include local corrosion, the existence of a reaction mixture, and a decrease in potential due to solution resistance, which causes errors in measuring the corrosion rate.

6.5.3 Electrochemical Impedance Spectroscopy

Electrochemical impedance spectroscopy (EIS) can be used to characterize the electrochemical behavior that occurs at the metal/solution interface if a passive layer forms on the surface of the workpiece. By using the electric element model, the electrochemical reactions that occur can be predicted. This method measures corrosion

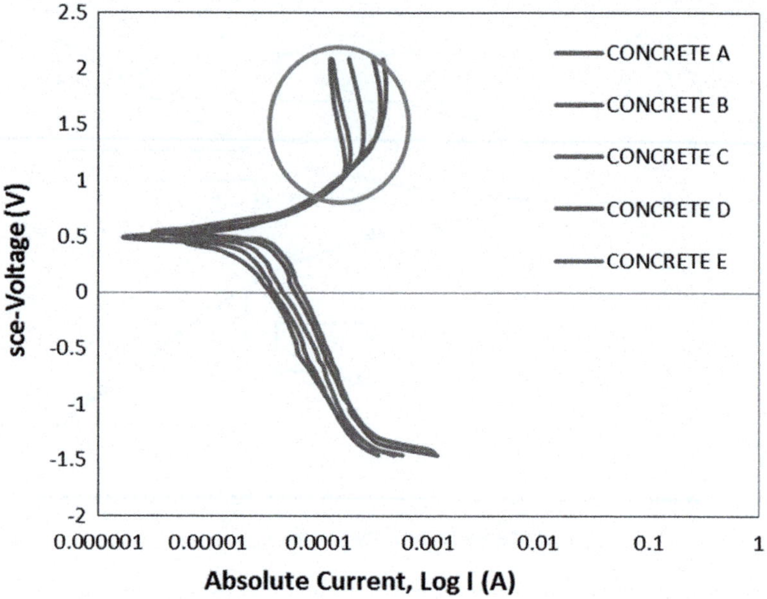

Fig. 6.10 The polarization of steel in various geopolymer concretes showing passive behavior [2]

rates by applying alternating current (AC) to an electrochemical cell and measuring the current non-destructively. The EIS instrument records the real (Z') and imaginary (Z") components when the workpiece is given an input frequency. The X-axis is the real value, and the imaginary value is recorded on the Y-axis, which is called the Nyquist plot. This method can also be used to study the electrochemical processes that occur on a metal surface when it reacts with an electrolyte. The relationship between the *X* and *Y* axes on the Nyquist plot is shown in Fig. 6.11 [15]. The relationship between impedance and corrosion rate is calculated using Eqs. 6.5–6.8.

$$Z(j\omega) = Z'(\omega) + jZ''(\omega) \tag{6.4}$$

The impedance value is computed using the following formula:

$$|Z| = \sqrt{|Z'|^2 + |Z''|^2} \tag{6.5}$$

with

$$\mathrm{Re}(Z) = |Z|\cos\phi \quad \text{and} \quad \mathrm{Im}(Z) = |Z|\sin\phi$$

where Z' is the real axis and Z'' is the imaginary axis. Hence,

$$\mathrm{R_p} = \text{Low frequency} - \text{High frequency} \tag{6.6}$$

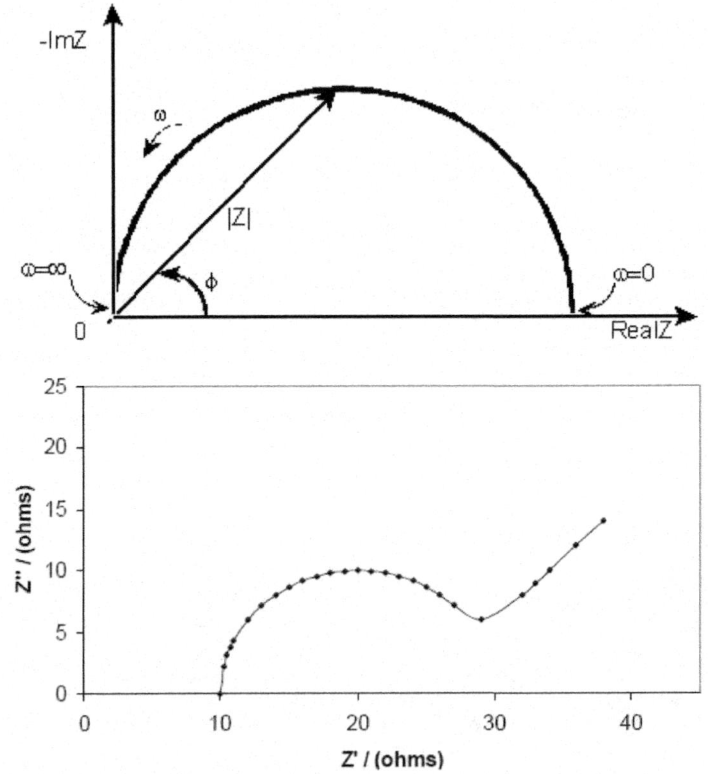

Fig. 6.11 The relationship between impedance magnitude (Z) and phase angle in the corrosion-charge transfer activation reaction (**a**) and the combination of activation and mass transfer reactions (**b**)

Corrosion speed is calculated using Eq. (6.7) as follow:

$$I_{\mathrm{corr}} = \frac{B}{Rp} \tag{6.7}$$

Measurement of steel in concrete using the EIS method uses the frequency applied to reinforcing steel embedded in concrete. The use of frequencies in steel produces various Nyquist plots (Figs. 6.11, 6.12 and 6.13). Furthermore, the Nyquist plot is read by analogy with an equivalent electric circuit consisting of resistance, inductance, and capacitance. The resistance consists of resistance by solution (R_s), resistance from charge transfer (R_c), phase constant element (CPE), and Warburg impedance (Fig. 6.12). In the cement cavity, the R_c shows that there is a resistance effect on charge transfer involved in the corrosion mechanism.

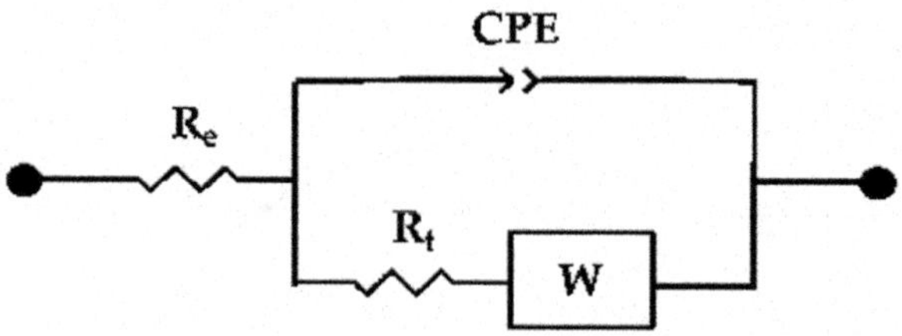

Fig. 6.12 Nyquist diagram for modeling corrosion in reinforcing steel where corrosion occurs due to activation and diffusion reactions. The flow of electrons that occurs in steel in concrete is modeled to occur due to resistance by solution (R_e), resistance due to charge transfer (R_t), resistance due to diffusion (W), and phase constant element (CPE)

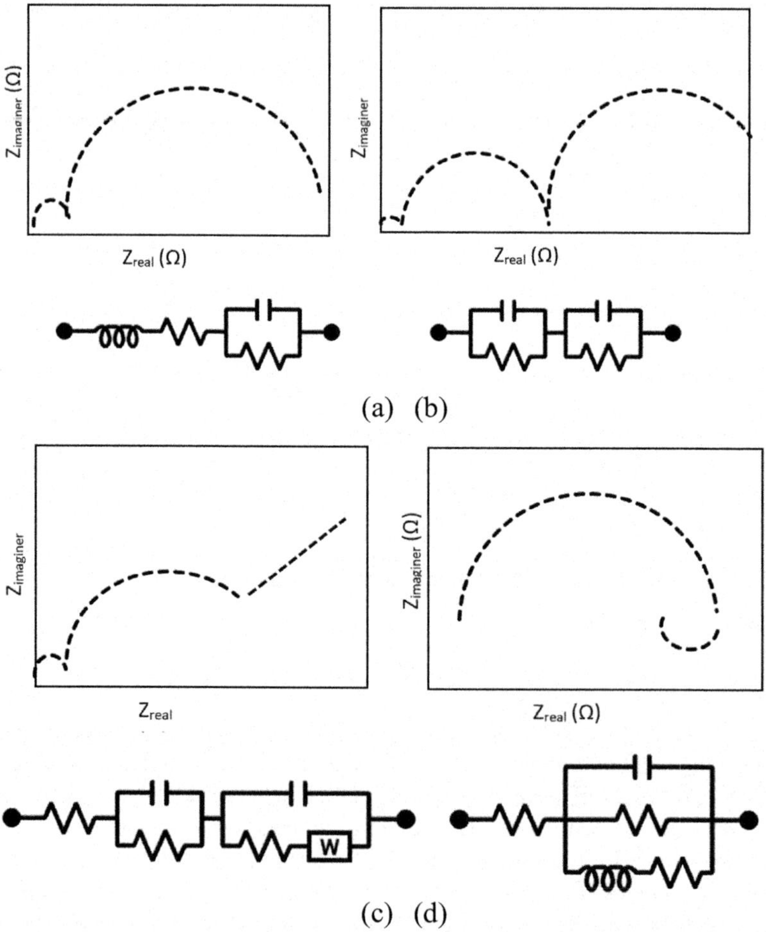

Fig. 6.13 Several models of corrosion mechanisms in steel reinforcement in concrete. The data from the EIS are used to calculate the corrosion rate of the steel by constructing a suitable analog circuit

6.6 Corrosion Potential Mapping (Corrosion Potential Survey)

Corrosion mapping is done by measuring the potential over a large area (see Fig. 6.14). Corrosion mapping is often carried out against corrosion attacks that occur in reinforced concrete. Corrosion is an electrochemical phenomenon whose potential can be measured by attaching a RE connected to a steel wire. The potential changes that occur depend on the corrosion activity. Corrosion maps provide detailed information about possible corrosion activity. General guidelines for identifying corrosion probabilities based on suggested half-cell potential values as in ASTM C876 are given in Table 6.3.

In measuring the corrosion potential in reinforced concrete, this technique should not be used separately, but the composition of the concrete must also be examined, namely the measurement of chloride content, pH, and concrete chemistry at a certain depth. From the corrosion potential data, the electrical resistance of concrete can be calculated. Electrical resistance plays an important role in determining the quality of concrete. This parameter is expressed in terms of resistivity in ohm-cm. Table 6.4 shows general guidelines for resistivity values based on which areas with a possible risk of corrosion can be identified in concrete.

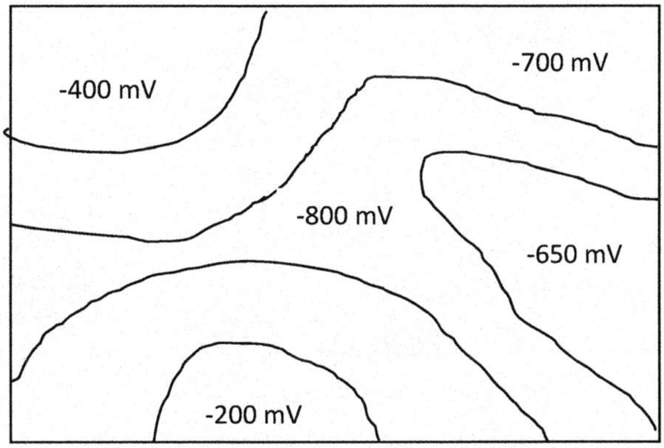

Fig. 6.14 An example of the measurement results of a corrosion potential map showing the risk of corrosion in a cross-sectional area

Table 6.3 Map of possible corrosion risks in reinforced concrete according to potential values based on ASTM C876-99 [5, 11]

SHE	CSE (mV)	Reinforced concrete corrosion rate
>120	>−200	Low (10% risk of corrosion)
120 to −30	−200 to −350	Intermediate corrosion risk
<−30	<−350	High (>90% risk of corrosion)
<−180	< −500	Severe corrosion

Table 6.4 Corrosion risk from resistivity [9]

Resistivity (ohm-cm)	Corrosion probability
More than 20,000	Ignored
10,000–20,000	Low
5,000–10,000	High
Less than 5,000	Very high

6.6.1 Attachment Strength Testing

The bond strength test is carried out by placing a concrete cylinder on a loading frame equipped with hydraulic jacks and load cells. After that, the reinforcing rod embedded in the cylinder is pulled out (i.e., pull-out test). Variation in the length of embedded reinforcement is treated as a modified variable. This test is usually carried out on specimens that are 28 days old by pulling the reinforcing steel embedded in the concrete cylinder (see Fig. 6.15).

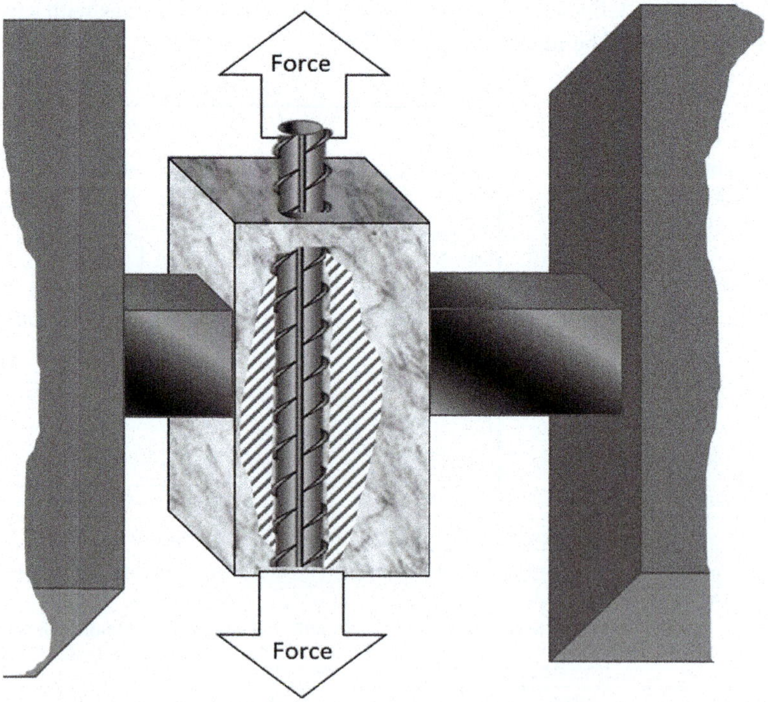

Fig. 6.15 Sketch of the bond strength test of steel to concrete (pull-out test)

6.6.2 *Penetration Testing of Chemical Elements in Concrete*

The quality of concrete is determined by the composition of the concrete. If the composition of the concrete changes, it indicates that there has been penetration of aggressive ions from the outside, which can damage the concrete and the reinforcement. An indirect way to determine the level of damage to reinforcement is by comparing the potential difference between reinforced concrete before and after contamination occurs (Fig. 6.16). Table 6.5 shows various test methods for chloride penetration in concrete.

The Fig. 6.17 shows the experiment for obtaining data on the penetration distance of aggressive elements that enter the concrete. The measurements are made by recording the potential difference between the electrodes at different locations and times in the concrete. The migration of chloride elements through the concrete is recorded by calculating the amount of chloride ion concentration present in the concrete and the NaOH solution. The NaCl solution is placed in the left chamber and another solution (usually NaOH) is placed in the right chamber with the two chambers separated by cement paste. Both solutions are saturated with $Ca(OH)_2$ to prevent a decrease in the amount of CaOH coming from the concrete.

Penetration testing of aggressive elements (CO_2, Cl, O_2) in concrete can also be carried out according to Fig. 6.18. The experiment is carried out by directly

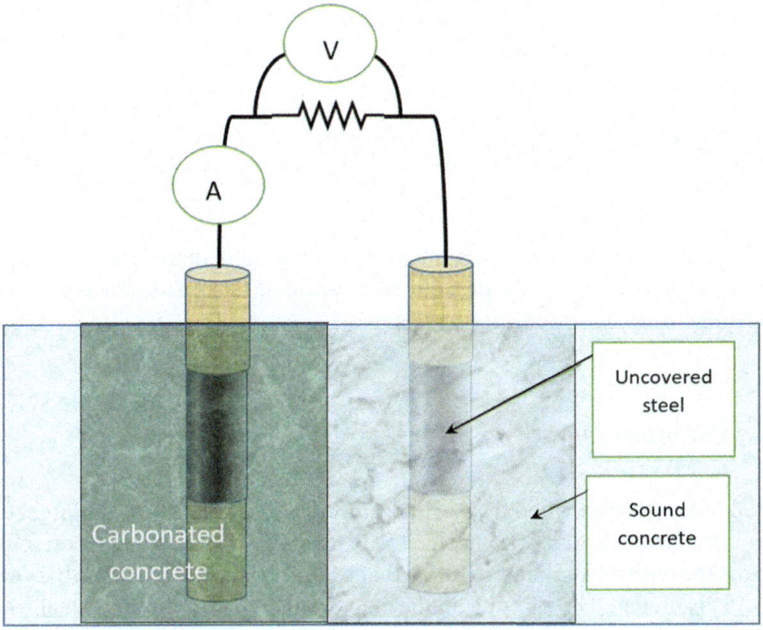

Fig. 6.16 Testing the effect of steel contaminated with carbon by comparing the potential value of steel before and after contamination

Table 6.5 Test methods for chloride penetration in concrete [8]

Category		Information	Test Methods
Long-term test (field conditions)		Chloride ion changes naturally under real conditions	• Bulk diffusion (ASTM 1556 [7]/NT BUILD 443) and chloride concentration profile (ASTM 1543/ AASHTO T260) [1]
Short-term test (accelerated recognition)	Conductivity test	Electric potential is used to accelerate the migration of chloride ions	• Rapid chloride permeability test (ASTM C1202 [4]) • Accelerated migration test (NT BUILD 492) • Chloride conductivity test (South African Durability Index Manual) • Electrical migration technique in which downstream chloride ion concentration is monitored
	Resistance test	Measures concrete resistance using a DC or AC current electric field	• Wenner four-probe method • Two-point method • Disk method • Surface-resistivity test FM 5–578 (FDOT 2004) • Concrete resistivity direct method (UNE 83,988–1) • UNE 83,987 multi-regime method
	Ionic change	Uses an electric field to record the movement of chlorine ions	• Whiting method • Permit method
	Pressure	Chloride ion penetration uses pressure on one side of the concrete	

incorporating aggressive elements into the concrete and simultaneously measuring the potential/current during the process. Measurements are made at various locations to determine the penetration speed of aggressive elements.

6.6.3 Potential Data Reading

Potential data obtained when measuring reinforced concrete are useful to determine the level of damage to reinforcing steel. The level of steel damage is determined by looking at the electrochemical reactions that occur in the steel. The Pourbaix diagram (Fig. 6.19) is prepared by plotting all the electrochemical reactions of steel when it is placed in various water conditions represented by pH and potential values. Concrete quality is assumed to have a correlation with the pH value. Besides being able to

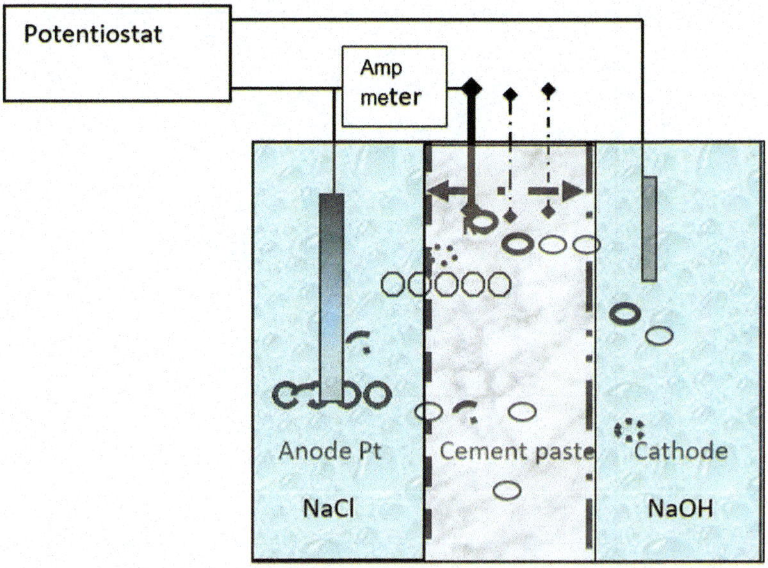

Fig. 6.17 Potential measurement between electrodes at different locations in concrete. Sodium chloride (NaCl) solution represents the destructive element and NaOH represents the solution present in the concrete

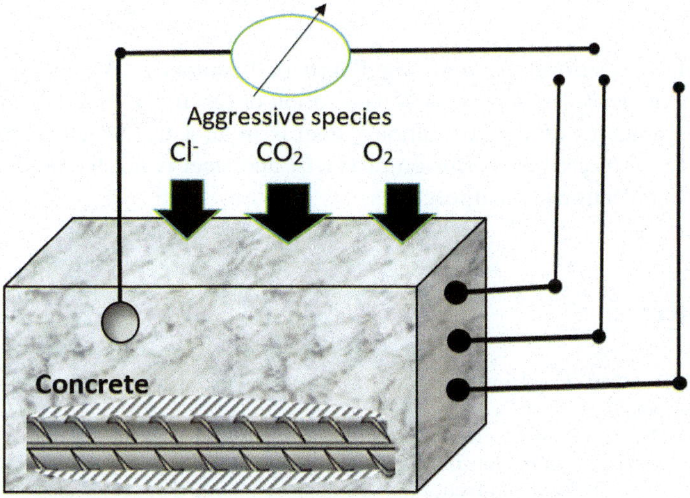

Fig. 6.18 Aggressive ion migration test scheme in concrete. The position of the anode electrode is changed to obtain the conductivity value of the concrete due to aggressive chemical elements

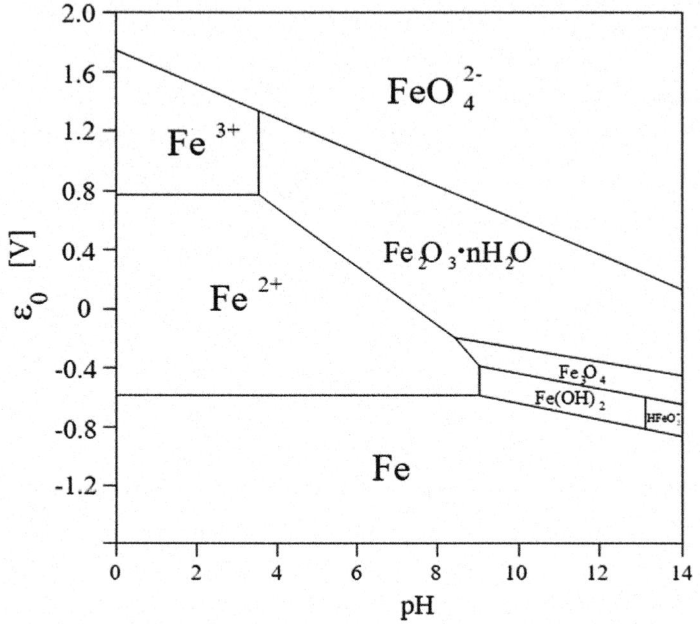

Fig. 6.19 Pourbaix diagram of steel in various water qualities showing the corrosion reaction products that occur when steel is present in various water qualities and potentials [15]

identify electrochemical reactions, the Pourbaix diagram can also identify reaction products that can be associated with passivation or reactivity. In the diagram, there are three categories of steel conditions: steel is in an active state, steel is passive because it is protected by corrosion products, and steel is not thermodynamically experiencing electrochemical reactions.

References

1. AASHTO T 260: Standard Method of Test for Sampling and Testing for Chloride Ion in Concrete and Concrete Raw Materials (1997)
2. Asmara, Y.P.: Long term corrosion experiment of steel rebar in fly ash-based geopolymer concrete in NaCl solution. Int. J. Corrosion Hindawi **1**, 1 (2016)
3. ASTM B117–19: Standard Practice for Operating Salt Spray (Fog) Apparatus
4. ASTM C1202: Rapid chloride permeability test, which is widely used, quickly determines how resistant your concrete is to chloride ingress (2022)
5. ASTM C876–15: Standard Test Method for Corrosion Potentials of Uncoated Reinforcing Steel in Concrete (2022)
6. ASTM E1508: Energy Dispersive Spectroscopy using an Electron Scanning Microscope SEM/EDS
7. ASTM-C1556: Standard Test Method for Determining the Apparent Chloride Diffusion Coefficient of Cementitious Mixtures by Bulk Diffusion (2022)

8. Dhana, B.S., Manu, S.: Comparative evaluation of various electrical methods used to assess chloride transport in concrete. Conference: FIB Congress 2014, Mumbai, Mumbai, January (2014)
9. Fontana, M.G.: Corrosion Engineering, 3rd edn. McGraw-Hill, New York (1986)
10. Hamidreza, Torbati, S.: Corrosion improvement of carbon steel in concrete environment through modification of steel microstructure. J. Mater. Civil Eng. **31**(5) (2019)
11. John, P., Broomfield: Corrosion Books: Corrosion of Steel in Concrete (2007)
12. Kelly, R.G., Scully, J.R., Shoesmith, D.W., Buchheit, R.G.: Electrochemical Techniques in Corrosion Science and Engineering. Marcel Dekker, New York (2003)
13. Mansfeld, F.: The polarization resistance technique for measuring corrosion currents. In: Fontana, M.G., Staehle, R.W. (Eds.), Advances in Corrosion Science and Technology. Plenum, New York (1976)
14. Scully, J.R.: Polarization Resistance Method for Determination of Instantaneous Corrosion Rates, Corrosion (Houston), Vol. 56, pp. 199–218 (2000)
15. Uhlig and Revie, R.W.: Corrosion and Corrosion Control: An Introduction to Corrosion Science and Engineering (3rd edition), Wiley, New York (1985)

Chapter 7
Reinforced Concrete Protection

Abstract Reinforced concrete protection is a critical aspect of ensuring the long-term durability and performance of concrete structures. This chapter explores various methods and techniques employed to protect reinforced concrete from corrosion and other forms of deterioration. Material selection plays a vital role in reinforced concrete protection. Reinforcement modification techniques, such as cathodic protection and sacrificial anode systems, are effective strategies for mitigating corrosion. These methods create a protective electrochemical environment that reduces the corrosion rate of the reinforcing steel. Concrete modification techniques, such as the use of geopolymer concrete and the application of inhibitors, offer alternative approaches to enhance the performance and longevity of reinforced concrete structures. Coatings applied to the concrete surface offer an additional layer of protection. Thickening the concrete cover, the layer of concrete surrounding the reinforcement, is another effective strategy for reinforced concrete protection. By employing these various methods and techniques, reinforced concrete structures can be adequately protected, ensuring their longevity, structural integrity, and resistance to corrosion.

7.1 Selection of Materials

Knowing the types of materials and their properties is useful for making the right choices in designing structures that exposes to specific environments. Understanding the material is an important step to determine the service life of the construction by calculating the corrosion rate. The corrosion rate of a material is also determined by its geometry, fabrication, minerals, finish, and composition. One of the important aspects of materials science is to analyze the failure of a product, structure or component that does not function properly. Figure 7.1 is displayed the types of materials used for construction purposes.

Y. P. Asmara, *Concrete Reinforcement Degradation and Rehabilitation*,
Engineering Materials, https://doi.org/10.1007/978-981-99-5933-4_7

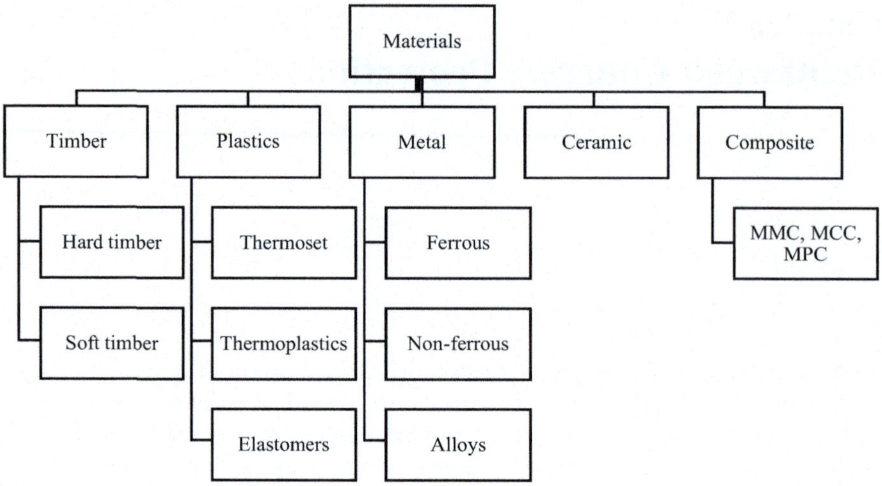

Fig. 7.1 Classification of materials based on their chemical composition [4] (*Ashby 2012*)

7.1.1 Carbon Steel

Mostly carbon steel is the first choice as a reinforcing material for improving tensile stress of concrete structures. Besides widely available in the market and cheap, carbon steel is also easy to fabricate and has allowable strength as a construction material. Although carbon steel is not corrosion resistant, but by using coating and treatment technology, corrosion can be minimized. In an environment polluted with CO_2 and H_2S gases, carbon steel is very weak to be attacked corrosion with the rate of up to 3 mm/year (Fig. 7.2) (*Asmara, 2009*). In a concrete environment, carbon steel has a corrosion rate that varies depending on the condition of the concrete (Fig. 7.3). The main characteristics of carbon steel are strong, ductile, shiny, solid, high thermal and electrical conductivity. It can form alloys with other metals, form ionic compounds and have a crystal structure. Carbon steels are classified based on the amount of carbon concentration, namely low, medium, and high carbon steels. Carbon steel can also be classified based on its microstructure, namely hypo-eutectoid steel ($C < 0.77\%$), hyper-eutectoid ($C > 0.77\%$), eutectoid ($C = 0.77\%$) and cast iron ($C > 2\%$) [4].

7.1.2 Stainless Steel (Stainless Steels)

Stainless steel has the main property of resistant to corrosion attack in various environments. The main alloying element is chromium with a concentration of about 11% by weight. The corrosion rate of stainless steel is very low in atmospheric environments. In polluted environments, stainless steels still have a low corrosion rate

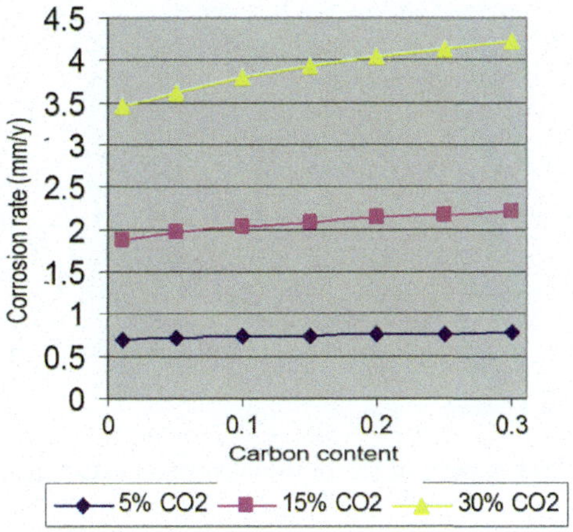

Fig. 7.2 Effect of carbon steel in various CO_2 environments on corrosion rates. (Calculated with ECE software [5]). **a** At 0% H_2S, total pressure 80 bar inlet and 30 bar outlet, 400 °C. **b** At 0% H_2S, total pressure 80 bar inlet, and 50 bar outlet, 400 °C

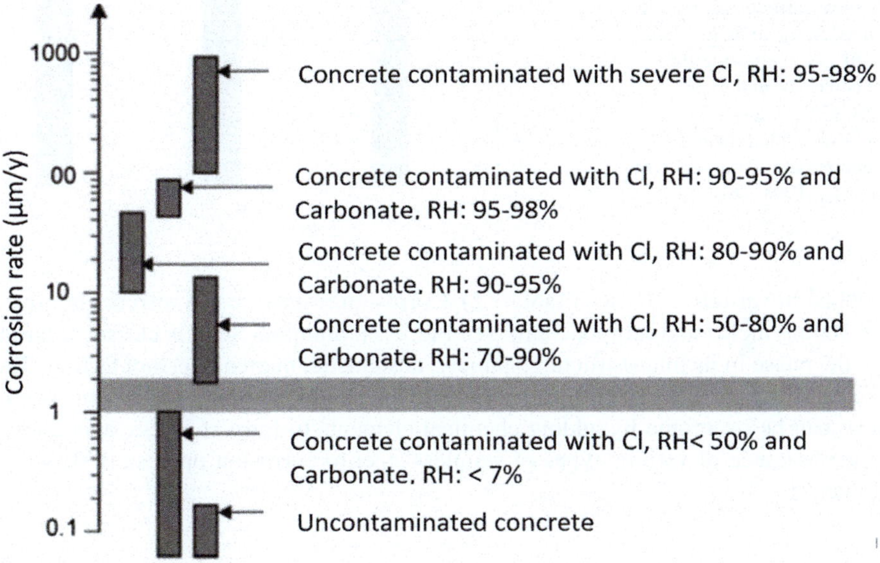

Fig. 7.3 Corrosion rate of exposed carbon steel in Cl and Carbonate contaminated concrete under various humidity conditions [3]

(Fig. 7.4). Corrosion performane of stainless steel as reinforcing concrete in chlorine environment is presented in Fig. 7.5. Figure 7.6 shows comparison corrosion properties of stainless and carbon steel. It can be seen that before corrosion process start, there is initiation periode which indicates reistivity of materials in withstand level of chlor ions. The use of stainless steel to strengthen concrete has been widely

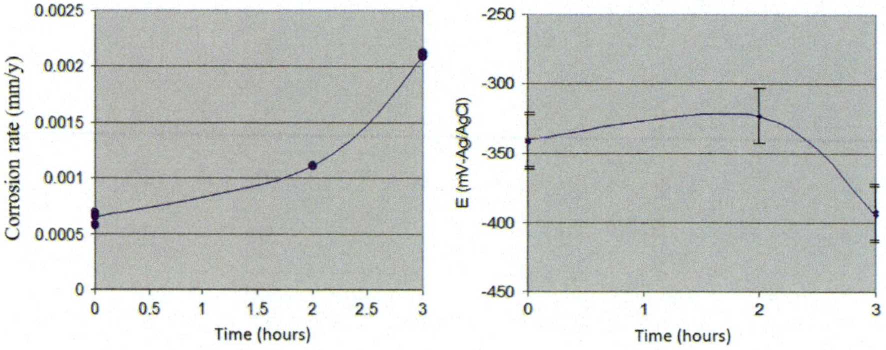

Fig. 7.4 Corrosion rate and changes in SS corrosion potential in CO_2 gas environment (*ECE*)

Fig. 7.5 The current density that occurs in carbon steel in concrete contaminated with 3% chloride is related to 3 types of materials. That is, passive carbon steel bars in chloride-free concrete (**a**), 316L SS in chloride-free concrete (**b**) and in 3% (20 °C, 95% RH) chloride-contaminated concrete (**c**). (The Concrete Society, 1998) [6]

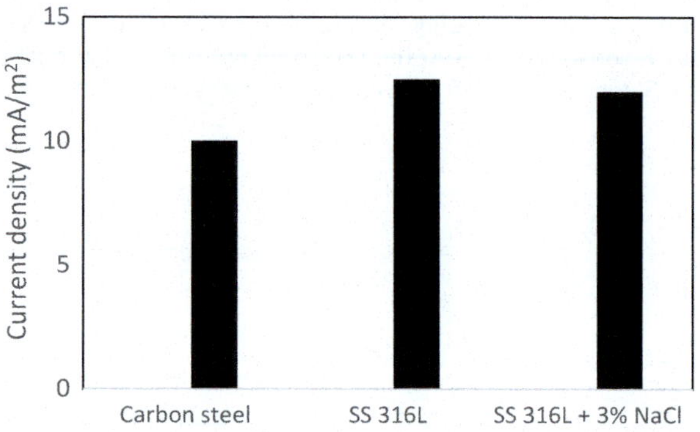

applied in various countries (Table 7.2). Corrosion resistance can also be improved by adding nickel and molybdenum (see Fig. 7.7). Stainless Steel is classified based on the phase in its microstructure, namely: austenistic, martensitic, and ferritic. The weakness of stainless steel in case of continuous and fluctuating loads. This weakness can be overcome by adding chemical elements to form stainless steel alloys. The resistance of various types of stainless steel to corrosion process is shown at Table 7.1.

7.2 Reinforcement Modification

7.2.1 Cathodic Protection

Cathodic protection (CP) is a technique to prevent metal corrosion by changing metal polarization. This method is carried out by applying a negatively charged current to the structure to be protected (see Fig. 7.8). When a structure is carrying a

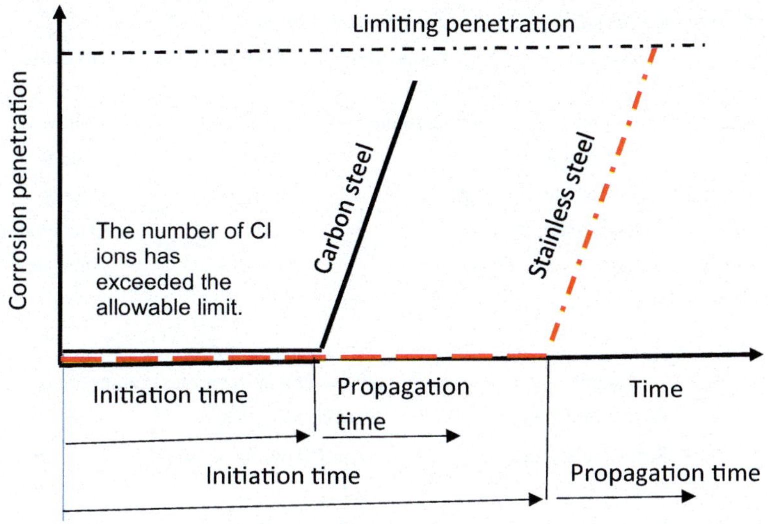

Fig. 7.6 Comparison of the use of carbon steel and stainless steel as concrete reinforcement against the speed of damage propagation bethone [6]

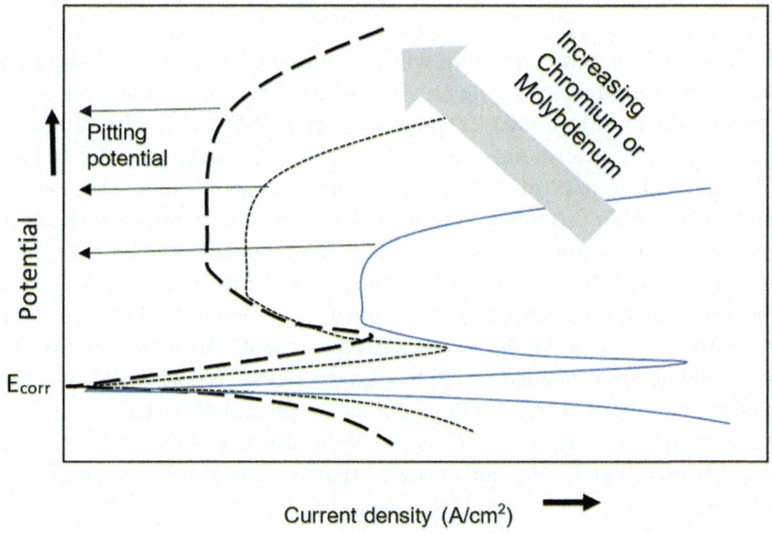

Fig. 7.7 Effect of chromium and molyvdenum composition on the polarization diagram. Either chromium or molybdenum causes pitting potential and the corrosion potential increases while the corrosion rate decreases [7]

Table 7.1 Properties of stainless steel under various conditions

SS alloy	Corrosion resistance	High temperature resistance	Welding ability	Hardening
Austenistic	Resistant	Very resistant	Very good	Cold work
Duplex	Very resistant	Low	Good	Cannot
Ferritic	Intermediate	High	Low	Cannot
Martensitic	Intermediate	Low	Low	Quenc/temper
Precipitation hardening	Intermediate	Low	Good	Age hardening

Table 7.2 The use of stainless steel as concrete reinforcement in various countries [6]

304	Bridge on I-696	Detroit, MI, USA	1984
304L	Schaffhausen bridge	River Rhine, Switzerland	1995
304LN	Guildhall	East London	2000
316	Underpasses	Newcastle, Tyneside, UK	1995
316L	Broadmeadow Bridge	Dublin, Ireland	2003
316LN	Gladstone Bridge	Queensland, Australia	

negative current, the entire structure will be cathodic. Thus the structure no longer releases electrons to the electrolyte (environment). Negative currents can be sourced from outside (electricity network) (impressed current-*ICCP*) or from other metals that are more active (sacrificial anode-*SA*) as presented schematically at Fig. 7.10. With this method, ions positive migrate towards the metal under the electric field and the concentration difference. Figure 7.9 shows a diagram polarization used to determine amount of supplied current required to produce completely protection of metals. Another possibility is that hydroxyl ions are formed in some of the concrete components, causing the growth of salt deposits at the cathode. These salts can coat the steel cathode and cause reduced current requirements. To achieve high efficiency, an extra metal anode is needed which has a more negative potential. So that there is a continuous supply of current to the metal being protected. Usually, it uses anodes that are inert (not corroded). It is also possible to use semi-inert electrodes (such as graphite, silicon alloyed iron), and corroded (consumable) anodes such as scrap iron [7, 14].

$$O_2 + 2H_2O + 4e- \rightarrow 4OH_- \quad (7.1)$$

If the supply of electrons is excessive, it will cause the formation of excess hydrogen gas (hydrogen evolution reaction).

$$2H_2O + 2e- \rightarrow 2OH_- + H_2 \quad (7.2)$$

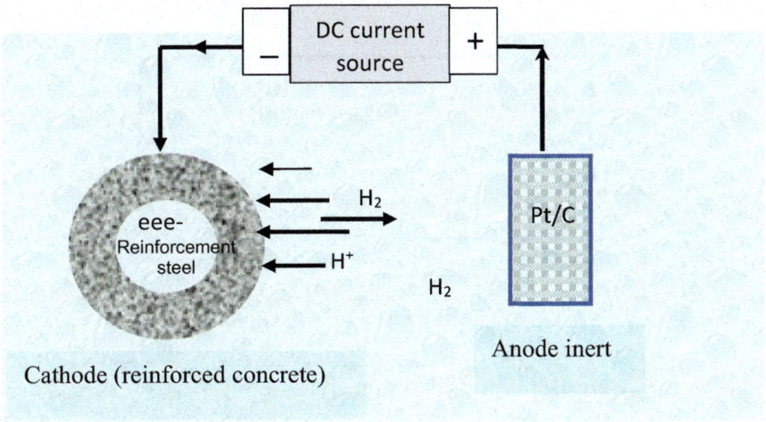

Fig. 7.8 Schematic of the circuit in the cathodic protection of the impressed current system [7, 14]

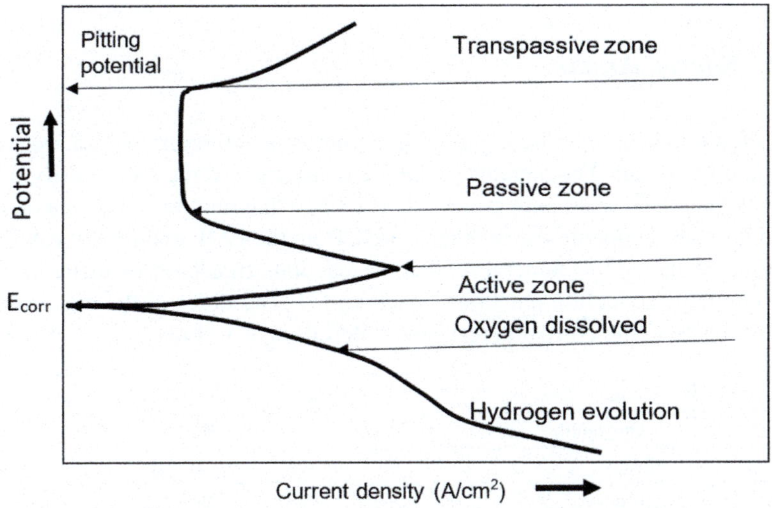

Fig. 7.9 A polarization diagram showing the properties of a metal when an electric current flows through it. Cathodic protection is carried out by applying current until the metal is in the cathodic zone where the hydrogen and oxygen reduction reactions occur. While anodic protection is carried out by bringing metal towards the passive zone

In the anode site, there is no actual anode reaction, because the anode is made of noble metals (inert), so the reaction that occurs is the electrolysis of water to release oxygen.

$$2H_2O \rightarrow O_2 + 4H^+ + 4e^- \quad (7.3)$$

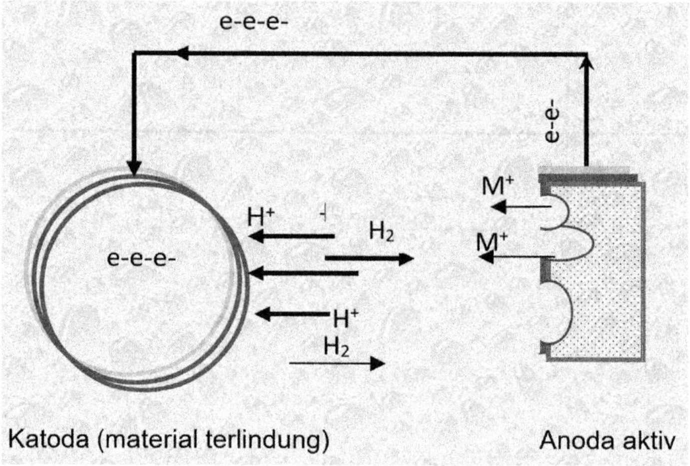

Fig. 7.10 Protection scheme with a sacrificial anode [7, 14]

7.2.2 Sacred Anode

The sacred anode is done by combining the protected structure with a more active (more anodic) metal. The metal is said to be more active if the metal easily releases electrons than the metal being protected. The EMF diagram provides data on reduction and oxidation potentials. In other words, from the EMF diagram, it can be seen which materials tend to undergo oxidation reactions (reactive) or materials which tend to undergo reduction reactions (inactive) [7]. The order of the materials from the most active to the corrosion resistant materials is as follows:

1	Zinc	7	Lead
2	Aluminum	8	Brass
3	Steel	9	Copper
4	Iron	10	Bronze
5	Nickel	11	Stainless Steel
6	Tin	12	Gold

7.3 The Effect of the Electric Field on the Chemical Distribution of Concrete

Cement as the main element of the concrete mix has the following composition: CaO (80–67%), Silica (17–25%), Al_2O_3 (3–8%), Fe_2O_3 (0.5–6%), MgO (0.1 –4%), SO_3 (1–3%). Hydration of NaOH and KOH forms Na_2O (0.15%) and will increase the pH, porosity, and value of a saturated $Ca(OH)_2$ solution.

Cathodic protection stops the anodic reaction of the steel and reduces the potential for corrosion in concrete. If corrosion stops, a cathodic reaction occurs involving the reduction of dissolved oxygen.

$$O_2 + 2H_2O + 4e- \rightarrow 4OH- \quad (7.4)$$

This salt can coat the steel cathode and cause reduced current requirements and if excessive cause reduction of more steel and will cause the hydrogen evolution reaction.

$$2H_2O + 2e- \rightarrow 2OH_- + H_2 \quad (7.5)$$

In the anode area, there is no actual anode reaction, because the anode is made of noble metals (inert), so the reaction that occurs is the electrolysis of water to release oxygen.

$$2H_2O \rightarrow O_2 + 4H + 4e- \quad (7.6)$$

The generated hydrogen ions cause the environment to become acidic and the current requirement increases.

In protection of reinforced concrete in seawater, chloride ions move towards the anode. These ions are released from the concrete. At the anode, chloride ions form chloride gas again. Hydroxide ions at the anode will also go to the anode or form $Ca(OH)_2$ compounds in the concrete matrix which then react to produce oxygen through the reaction.

$$2OH- \rightarrow H_2O + 1/2O_2 + 2e^- \quad (7.7)$$

Minimum Concrete Cover Requirements [2].

7.4 Concrete Modification

7.4.1 Geopolymer Concrete

Combining concrete with polymers is an effort to produce corrosion-resistant concrete. Geopolymers have advantages such as: good tensile strength, light weight, high corrosion resistance and long service life. Therefore, in recent years, geopolymer has become a potential alternative to replace conventional portland cement concrete (OPC) used in infrastructure development (*Hafiz, 2017*). The strength of geopolymer concrete can be modified to get the best properties [10, 13], using a combination of fly ash and cement to improve the mechanics of geopolymer concrete. They concluded that geopolymer concrete has a prospective material that can be used as an alternative structural material to replace OPC. Further investigation, it was found that the main chemical in fly ash that contributes to increasing the compressive stress is calcium compounds (CaO and $Ca(OH)_2$). The geoplymer concrete compressive stress in the experiment reached 29.2 MPa with the addition of 3% CaO and 3% $Ca(OH)_2$.

7.4.2 Inhibitors

Inhibitors are chemical substances added to concrete in small concentrations to delay the time for corrosion to occur in the concrete structure. Corrosion inhibitor mixtures work by increasing the passivation of steel reinforcement. This method can resist corrosion if the passivation in the concrete formed. However, passivation on steel will be lost due to the entry of chloride or carbonation. Inhibitors are added to concrete during the manufacture of concrete. This method can significantly reduce the cost of maintaining reinforced concrete structures. Some of the most popular Corrosion inhibiting mixtures are, carboxylic amine, organic amine-ester emulsion, calcium nitrite, organic alkenyl dicarboxylic acid salt.

7.4.3 Green Inhibitors for Concrete

Green inhibitors are inhibitors made from naturally soluble materials. Green inhibitor ingredients are made from organic elements. Several organic materials that can be used as green inhibitor ingredients have been studied by several researchers as described below:

a. Quraishi [12] in his research, he studied the effect of calcium palmitate combined with calcium nitrite on the corrosion of steel in concrete. The results showed that calcium palmitate was an effective inhibitor. The inhibitors provide 91% to 92% efficiency after 90 days of exposure time in 3.5% NaCl solution. Inhibitors also showed no effect on the mechanical strength of cement and concrete. Petrographic examination revealed that calcium palmitate clogged the pores and reduced the corrosion rate of the steel. Further investigations found that calcium palmitate inhibited corrosion through an adsorption mechanism. The inhibitor creates a film onto the steel surface via polar carboxylic groups that block the pores forming an insoluble hydrophobic iron stearate salt.

b. Quraishi [12] investigated the performance of calcium stearate as an inhibitor. They embed carbon steel on OPCIS:456–2000. The results showed that the efficiency of the inhibitor was achieved at 90% and 93% at a concentration of 3% and 5% in 60 days of experiment using 3.5% NaCl, respectively. Like previous studies, these inhibitors reduce the corrosion rate of steel by blocking porous concrete to limit the ingress of chloride ions.
c. Joshua [9] characterized phyllanthus muellerianus as an inhibitor to reduce corrosion of steel reinforcement in industrial environments. They used 0.5 M H_2SO_4 to simulate an industrial/microbial environment. At a concentration of 6.67 g/l, this inhibitor reduced the corrosion rate of reinforcement up to 90%. Meanwhile, at a concentration of 1.67%, the reduction corrosion rate was 78%. From the research conducted, the leaves of phyllanthus muellerianus and euphorbiaceae contain tannins, phlatanins, saponins, flavooids, terpenoids and alkaloids.
d. Abbas [1] investigated the effectiveness of green inhibitors extracted from orange peel waste. They extracted dried orange peel using methanol extract at 6 h immersion time in methanol at pressure (60 mbar) and 40 °C. From the experimental data using electrochemical polarization measurements and weight loss testing for 7 days of immersion time, it showed performance which was good in 3.5% NaCl solution. The corrosion rate of reinforcing steel decreased to 0.02 mm/year at 3% inhibitor concentration.

7.5 Coating

Coating is one method of metals treatment to avoid direct contact between metal and environemts. Surface treatment is a process that change the properties of a material's surface, such as its texture, adhesion, corrosion resistance, and appearance. The purpose of surface treatment can vary depending on the application, but typically it is done to improve the surface's performance or aesthetic qualities. Figure 7.11 shows various of surface treatment applied to protect metals. Coatings are also used to maintain the aesthetics of structures and extend its service life. The coating system uses a solid material that can stick to the metal surface through either a physical or chemical reaction. The coating plays a role in insulating the direct contact between the metal surface and the environment to prevent the corrosion process takes place. Coatings can be classified based on their function, namely [7].

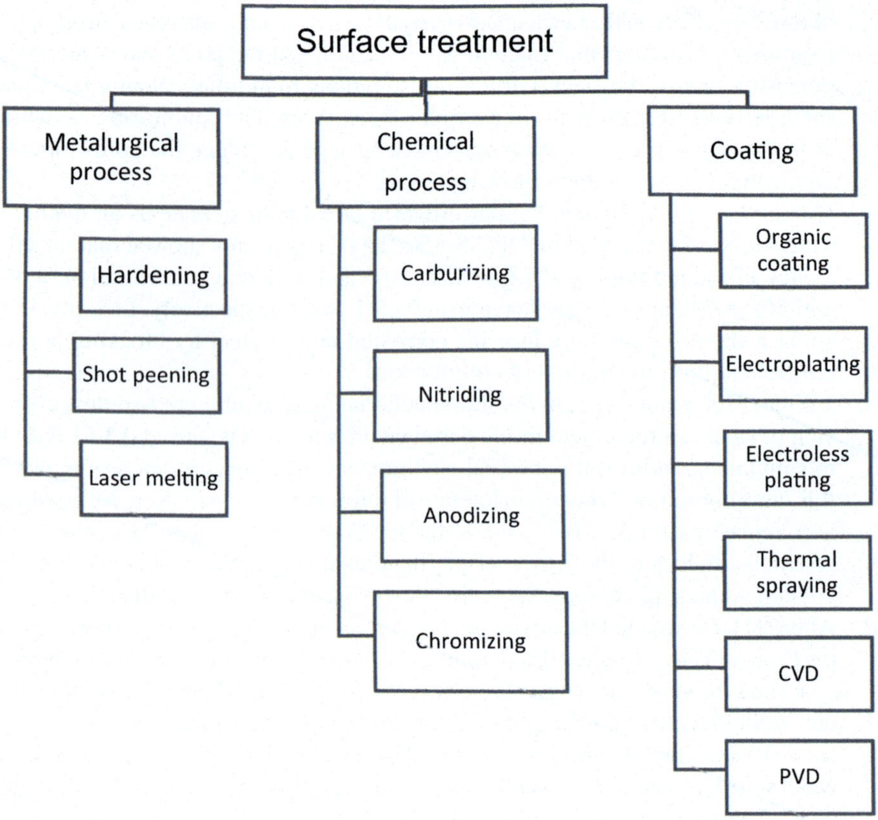

Fig. 7.11 Types of surface treatment processes to improve material quality [15]

- Insulation; i.e., provide protection by preventing water, oxygen, and electrolyte contact with metal surfaces.
- Inhibitive; protection that relies on chemicals that function to inhibit corrosion reactions by changing the electrolyte solution.
- Sacrifice: a thin metal layer which has a lower electrode potential value or which has a higher level of activity than the metal being protected.
- A combination of insulation, inhibition, and sacrifice.

Materials typically used in protective coatings are: polymers, epoxy, and polyurethane for non-metallic coatings. Zinc, aluminum, and chromium for metallic coatings. Techniques applied are: sprayed, welding, metallized, swept. Factors that be considered before doing the coating are: surface preparation, applying primer, and select sealant. However, the application of coating can be ineffective which lead to fail formation of protective layer due to:

- Mechanical impact

It caused by scratching, friction, or erosion of solid particles.

- Incompatibility

It happens when the coefficient of thermal expansion between the metal surface and the protective layer itself.

- Chemical effects

It happens when organic solvents or strong oxidants occur in an active environment.

- Solar radiation

The UV spectrum of solar radiation causes polymer chain disruption or oxidation.

7.5.1 Thermal Spraying

Thermal spraying is the process of coating metal using metal that is sprayed onto the surface of the workpiece. To make metal sprayable, it needs a nozzle equipped with oxy fuel which can melt metal (see Fig. 7.12). The high pressure gas pressurze molten metal to form deposits a layer over the surface of the work piece. This process resembles the welding process.

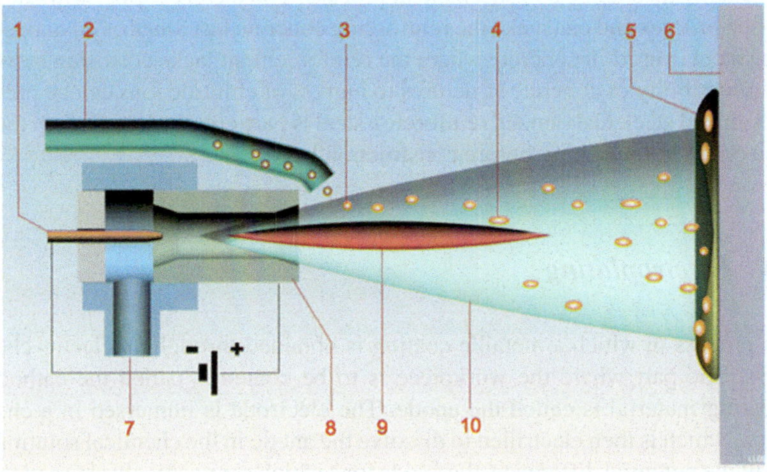

Fig. 7.12 Schematic cross-section of a typical atmosphere plasma spray process.1 Cathode. 2 Powder supply. 3 Powder particles. 4 Melted powder particles. 5 Splat 6 Substraat. 7 Fuel gas supply. 8 Anode. 9 Plasma arc. 10 Flame with melted powder particles. *Sources LaurensvanLieshout, CC BY-SA 3.0* <https://creativecommons.org/licenses/by-sa/3.0>, via *Wikimedia Commons*

7.5.2 Vapor Deposition

In this process, the material used for coating is made in a gaseous state during coating so that a very thin layer is produced. To make the material into the vapor phase, a vacuum chamber is needed at high temperature and very low pressure. The coating material is not only metal, but can also use several materials: carbide, nitride, or ceramic. This process can be applied to make equipment that has specific uses, namely surface hardening: cutting tools, drills, reamers. An example of vapor deposition is sputtering which is often used to make electronic chips. The suptering method uses an electric field which ionizes an inert gas (e.g. Argon), then the ions are used to shoot the coating material. The atoms of the coating material (cathode), when exposed to ions causes a change in the atomic lattice which then coats the workpiece through a condensation process. Aluminum sputtering examples are used to make most of the chip internal connections in the semiconductor industry.

7.5.3 Galvanization

Galvanized pipe can be used to protect concrete reinforcement. Galvanized steel has anti-rust properties. The metal coating used is a material that has more noble properties, such as copper on steel. This type of protective coating is only effective if the coating is free of pores or damage. Metallic coatings are applied using a sprayer, electrochemical, chemical or mechanical process. The metal coating slows down the corrosion process and can make the reinforcing concrete last longer. Galvanized reinforcing steel is used in concrete where the reinforcement lacks corrosion resistance. The susceptibility of concrete structures to ingress of chloride ions can be prevented by galvanized steel. Galvanized reinforcing steel is particularly useful when the rebar will be exposed to weather before construction begins.

7.5.4 Electroplating

It Is a process in which a metallic coating is obtained through an electro-chemical process. The part where the workpiece is to be coated is called the cathode and the coating material is called the anode. The electrode is immersed in a chemical solution, which is then electrified to dissolve the anode in the chemical solution. As a result of the potential difference, the anode ions which are positively charged towards the cathode (negatively charged) form a layer on the surface of the workpiece. An example of this process is copper plating using $CuSO_4$ solution, where copper acts as an anode. Figure 7.13 below is presented schematic principle of electroplating process.

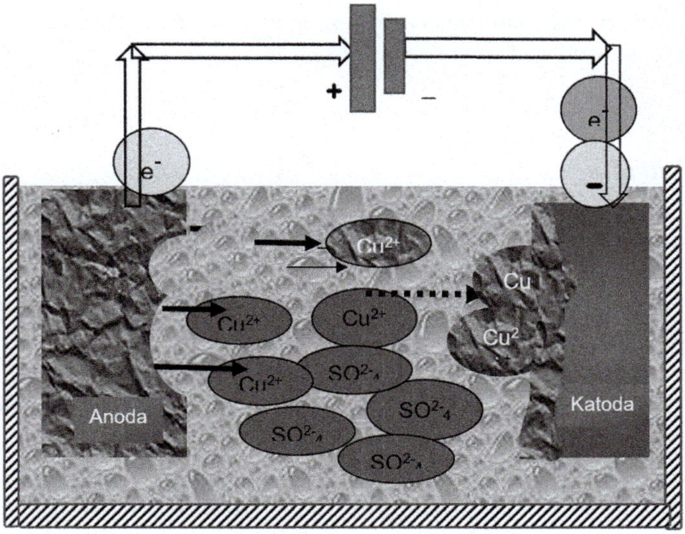

Fig. 7.13 The process of plating metal with copper (Cu) uses a solution of $CuSO_4$ [7, 8]

7.5.5 Electroless Plating

This process uses only chemical solutions to coat the metal. So that the coating process occurs as a result of chemical reaction between the coating material and the work piece. The use of this process is for coating metals with non-conducting materials. The most common use of this process is Nickel plating. Where nickel chloride undergoes reduction in sodium hypophosphite solution. Nickel plating aims for surface hardening. Electron nickel plating (ENP) plating is done by reducing nickel ions to nickel metal with a chemical reducing agent such as sodium hydrophosphite. The ENP improves resistance to corrosion attack caused by salt water, carbon dioxide, oxygen, and hydrogen sulfide. The use of high phosphorus (10–14%) produces an amorphous surface thereby reducing the number of grain or phase boundaries. ENP produces a uniformly thick coating over the entire workpiece. Without heat treatment, high phosphorus ENP still provides good hardness and wear resistance properties.

7.5.6 Anodizing

Anodizing is the process of making metal undergo an anodic reaction. A common example is aluminum plating. Where a layer on the surface of aluminum is formed due to an anodic reaction in a chemical solution. This process is carried out by making metal as the anode, an electrolytic reaction will occur which forms a hard metal oxide layer of the workpiece (anode). Examples of uses for this process include:

photo frames, car body parts, knobs, bathroom fixtures and shelves, sporting goods, such as baseball bats, and so on.

7.5.7 *Nitriding*

Nitriding is a type of treatment heat using nitrogen as a coating material. As a result, nitrogen deposits on the surface of the metal which causes the metal to harden. This process can be used for hardening steel, titanium, aluminum and molybdenum. An example of using nitriding is hardening gears, axis crank, valve, extruder, tool casting, printforge, extrusion dead, injector and tool sprint plastic. Nitriding can be done using nitrogenous axle. Nitriding is performed in a large bath filled with a nitrogen-producing solution such as cyanide salt.

7.6 Thicken the Concrete Cover

Protecting reinforcing steel in concrete from corrosion attack can also be done by increasing the thickness of the concrete. The recommended minimum thickness of concrete to resist corrosion is shown in Table 7.3. Concrete is a non-conductive and tough material that is resistant to penetration by outside elements/particles. Creating a thick layer covering the reinforcing steel allows for increased resistance and minimizes ion diffusion in the concrete.

Table 7.3 The minimum thickness recommended to withstand the corrosion rate in reinforced concrete [2, 11]

Types of concrete	Min (mm)
Concrete cast against and permanently exposed to earthConcrete exposed to earth or weather: No. 19 (No. 6) through No. 57 (No. 18) bars No. 16 (No. 5) bar, MW200 (W31) or MD200 (D31) wire, and smaller Concrete not exposed to weather or in contact with ground: Slabs, Walls, Joists: No. 43 (No. 14) and No. 57 (No. 18) bars No. 36 (No. 11) bars and smaller Beams, columns: Primary reinforcement, ties, stirrups, spiralsShells, folded plate members: No. 19 (No. 6) bar and larger No. 16 (No. 5) bar, MW200 (W31) or MD200 (D31) wire, and smaller	75 50 40 40 20 40 20 15

(continued)

Table 7.3 (continued)

Types of concrete	Min (mm)
Precast Concrete1	
Concrete exposed to earth orweather: wall panels: No. 43 (No. 14) and No. 57 (No. 18) bars No. 36 (No. 11) bars and smaller Other members: No. 43 (No. 14) and No. 57 (No. 18) bars No. 19 (No. 6) through No. 36 (No. 11) bars No. 16 (No. 5), MW200 (W31) or MD200 (D31) wire, and smaller Concrete not exposed to weather or in contact with ground: Slabs, walls, joists: No. 43 (No. 14) and No. 57 (No. 18) bars No. 36 (No. 11) bars and smaller Beams, columns: Primary reinforcement Ties, stirrups, spirals Shells, folded plate members: No. 19 (No. 6) bar and larger No. 16 (No. 5) bar, MW200 (W31) or MD200 (D31) wire, and smaller	40 20 50 40 30 30 15 d_p momtnot tutorings than 15 and ned not exceed 4010 15 10
Prestressed Concrete2	
Concrete cast against and permanently exposed to earthConcrete exposed to earth or weather: wall panels, slabs, joistsOther members Concrete not exposed to weather or in contact with ground: Slabs, walls, joists Beams, columns: Primary reinforcement Ties, stirrups, spirals Shells, folded plate members: No. 16 (No. 5) bar, MW200 (W31) or MD200 (D31) wire, and smaller other reinforcement	75 25 40 20 40 25 10 d_b momtnot tutorings than 20

1 Manufactured under plant-controlled conditions.
2 Modification to the cover requirements is possible depending on the manufacturing method and tensile stress in the member. See [2].
db = diameter of reinforcing bar

References

1. Abbas, A., Török: Corrosion studies of steel rebar samples in neutral sodium chloride solution also in the presence of a bio-based (green) inhibitor. Int. J. Corros. Scale Inhib. **7**(1), 38–47 (2018)
2. ACI 318–02: American Concrete Institute, Farmington Hills, Michigan (2002)
3. Bertolini, L., Elsener, B., Pedeferri, Polder: Corrosion of Steel in Concrete: Prevention, Diagnosis, Repair. Weinheim, Germany: Wiley-VCH Verlag GmbH & Co. (2004)
4. Callister, D., David, G., Rethwisch: Materials Science and Engineering an introduction. Department of Metallurgical Engineering, The University of Utah (2017)

5. ECE, Electronic Corrosion Engineer, Intetech. ECE 4th Edition, Corrosion Prediction Model. https://www.woodplc.com/solutions/expertise/az-list-of-our-expertise/software-and-products/ece-corrosion-and-materials-selection
6. Federica, L., Maddalena, C., Matteo, Elena, R.: Corrosion behavior of stainless steel reinforcement in concrete. Corrosion Rev. **37**(1), 3–19 (2019)
7. Fontana, M.G.: Corrosion Engineering, 3rd edn. McGraw-Hill, New York (1986)
8. Jones: Principles and Prevention of Corrosion (2nd edition), Prentice Hall, Upper Saddle River NJ (1996)
9. Joshua, O.A., Olubanke: Investigating prospects of Phyllanthus muellerianus as ecofriendly/sustainable material for reducing concrete steel reinforcement corrosion in industrial/microbial environment. Energy Procedia. **74**, 1274–1281 (2015)
10. McLellan, B.C., Williams, R.P., Lay, A. van Riessen, Corder: Costs and carbon emissions for geopolymer pastes in comparison to ordinary Portland cement. J. Cleaner Prod. **19**(9–10), 1080–1090 (2011)
11. PCA: Concrete Slab Surface Defects: Causes, Prevention, Repair, IS177, Portland Cement Association (2001)
12. Quraishi, M., Abhilash, P.P., Singh: Calcium stearate: a green corrosion inhibitor for steel in concrete environment. J. Mater. Environment. Sci. **2**(4), 365–372 (2011), ISSN: 2028-2508
13. Rashad, A.M.: Potential use of phosphogypsum in alkali-activated fly ash under the effects of elevated temperatures and thermal shock cycles. J. Clean. Prod. **87**(1), 717–725 (2015)
14. Uhlig, Revie, R.W.: Corrosion and Corrosion Control: An Introduction to Corrosion Science and Engineering (3rd edition), Wiley, New York (1985)
15. Zia Ullah, A.: Effect of spraying parameters on surface roughness, deposition efficiency, and microstructure of electric arc sprayed brass coating. Int. J. Adv. Appl. Sci. **7**(7), 25–39 (2020)

Chapter 8
Concrete Reinforcement Inhibitors

Abstract In this chapter, it discusses the crucial role played by corrosion inhibitors and their significant influence on the longevity and safeguarding of concrete structures. The chapter extensively explores the concept of inhibitors, giving particular emphasis to the process of selecting the most suitable ones and categorizing them based on their protective methods. Furthermore, practical aspects pertaining to their application are discussed, emphasize on how they are used in various construction practices. The distinction between inorganic and green inhibitors is also addressed, with a special focus on environmentally friendly alternatives that effectively preserve the integrity of steel reinforcement. Additionally, the chapter extensively investigates the essential procedure of inhibitor testing and assesses their effectiveness in corrosion prevention. By conducting a comprehensive exploration of the complex interplay between inhibitors and the overall performance of concrete reinforcement, this chapter offers valuable insights that are indispensable for ensuring the enduring sustainability and durability of construction projects.

8.1 Inhibitors

Preventing corrosion that occurs in reinforcement in concrete is an essential method to extend service life of the structural building. Preventing corrosion of steel reinforcement cannot be done by relying solely on the passive layer contained in the concrete. The passive layer of concrete tends to be weak if chemical elements penetrate to the concrete layer [11]. Thus, it is important to strengthen passive layer formed on the steel surface. Modification of the surface of steel reinforcement becomes passive can be done by adding chemicals to the concrete, which are called inhibitors. Inhibitor is chemical which can reduce the rate of corrosion. Inhibitors are one of the easiest methods to reduce the corrosion rate. Inhibitors are applied to the concrete through several methods: mixing inhibitors while the concrete is in the molding or adding inhibitors after the concrete is finished. Inserting the inhibitor into the concrete pores can be done manually or electrochemically (see Fig. 8.1) inhibitors are needed in large quantities and should be continuously injected. This method has a negative impact on the environment. To avoid environmental problems, a new type of inhibitor has

Y. P. Asmara, *Concrete Reinforcement Degradation and Rehabilitation*,
Engineering Materials, https://doi.org/10.1007/978-981-99-5933-4_8

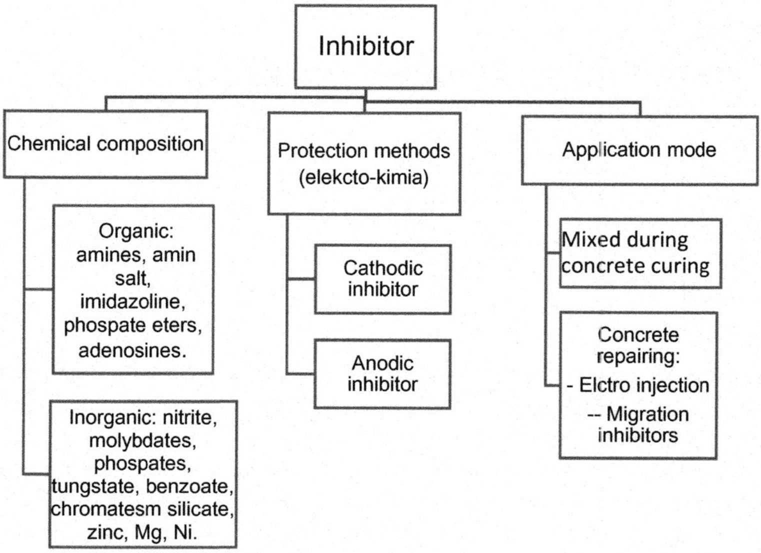

Fig. 8.1 Classification of inhibitors is divided into raw materials for making inhibitors; organic and inorganic, methods of protection: anodic and cathodic, and how to use them; used by mixing it when casting, used when the concrete is ready or for repairing damaged concrete [12]

been developed, namely inhibitors produced from plant extracts (green inhibitors). This type of inhibitor is friendly to the environment and can be used for all types of concrete besides economically affordable. Mechanism of inhibitor in reducing the corrosion process occurs by:

- Change the steel potential to be more positive. Or change the anodic/cathodic reaction based on the Tafel curve (see Fig. 8.2).
- Reducing the movement of atoms or inhibiting the process of ion diffusion to the metal surface.
- Increases surface resistance by creating a thin layer on the surface of the workpiece.

8.2 Selection of Inhibitors

Inhibitors have been mass-produced by industry and have been widely used to prevent corrosion. Many inhibitors are available on the market with special specifications. Knowledge of corrosion processes is required to choose the right type of inhibitor. Selection of the correct inhibitor is determined by the type of working conditions in which the inhibitor will be used. Inhibitor can be applied to protect metals in the outside air, in a closed system or in a circulating system. Temperature and pressure must be considered when choosing an inhibitor. Higher temperatures and pressures

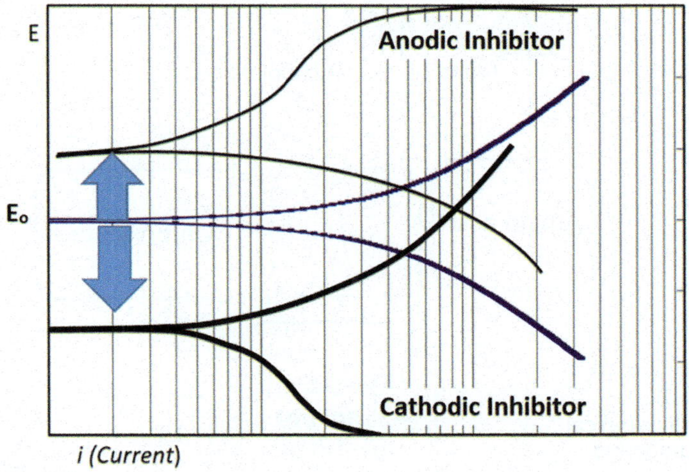

Fig. 8.2 How anodic protection and cathodic inhibitors work. Anodic inhibitor: reduces the rate of corrosion by changing the polarization towards the anode by forming a film layer at the anode. Cathodic inhibitor: changes the polarization to the cathode region

can lead to polymerization and sludge formation. The level of temperature and pressure will determine the corrosivity of aggressive chemical elements. The thermal stability of the inhibitor must be evaluated during the corrosion prevention process. Other factors that must be considered in the selection of inhibitors are: determining the problem to be solved, identifying corrosive elements, pressure, ambient temperature, velocity, pH, chemical composition, water/oil ratio, water salinity, and water and oil acidity. These conditions determine how much and what type of inhibitor should be added to inhibit the desired corrosion rate.

8.2.1 Classification of Inhibitors Based on the Method of Protection

Based on the protection mechanism, there are three types of corrosion inhibitors. The first, inhibitors that work by polarizing the corrosion potential in the anodic region are called anodic inhibitors (electrochemically). Passive anodic inhibitors by forming a film on the metal surface to reduce the corrosion rate. Examples of anodic inhibitors are chromates, nitrates, molybdates and tungstate. Second, cathodic inhibitors are inhibitors that work by polarizing the corrosion potential to the cathodic region. Basically, they limit the corrosive species through creating barrier in process of diffusion. The latter is a corrosion inhibitor that works by combining anodic and cathodic mechanisms. Corrosion inhibitors can also be classified as chemical or natural and organic or inorganic (see Fig. 8.3). Generally, inorganic inhibitors have cathodic and anodic polarization mechanisms [12].

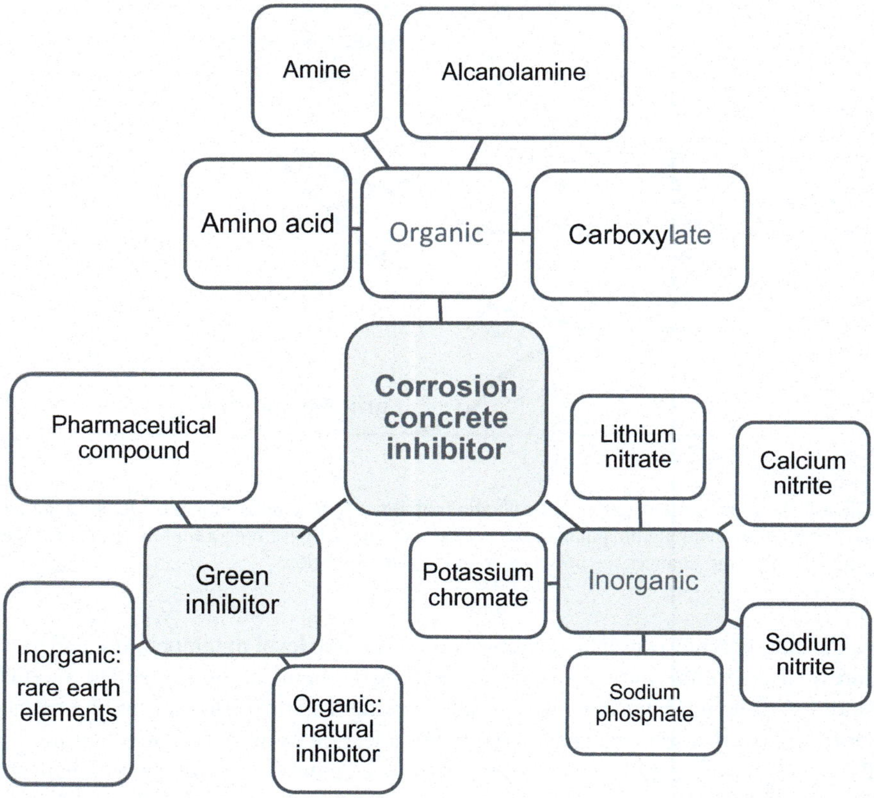

Fig. 8.3 Classification of inhibitors based on the ingredients mixed. These inhibitors are used to protect concrete reinforcement from corrosion attack [17]

8.2.2 Method of Applications

The first requirement before applying an inhibitor is to determine the amount of inhibitor to be used. The initial amount of inhibitor must be sufficient to cover the entire exposed metal surface. The second requirement is that if any part of the inhibitor is peeled off, a new layer of inhibitor should immediately reform. In reinforced concrete, inhibitors are introduced into the concrete in three ways. There ae conducted by mixed during concrete casting, injected electrochemically, and through methods of *migrating corrosion inhibitor (MCI).* The MCI is applying the inhibitor into the concrete in liquid phase through the capillary principle which causes the inhibitor to migrate in the vapor phase throughout the concrete's pore structure. When MCI comes into contact with the reinforcing steel, it is energizedionic attraction and forms a protective molecular layer. This film prevents the corrosive elements from further reacting with the reinforcement and also reduces the existing corrosion rate. This method has shown a greatly extending the service life of the concrete. The inhibitors

Table 8.1 Performance of inorganic inhibitors on the corrosion rate of steel in concrete contaminated with Cl ions [12, 22]

Systems	Corrosion rate (mmpy)		
	1%	2%	3%
Control	0.0093	–	–
$NaNO_2$	0.0023	0.0093	0.0250
ZnO	0.0022	0.0023	0.0024
Mixed	0.0009	0.0012	0.0037
Triethanolamine	0.0117	0.0038	0.0071
Monoethanolamine	0.0171	0.0042	0.0096
Diethanolamine	0.0203	0.0034	0.0083

used are made from amine alcohols and amine carboxylates. Amine inhibitors work to protect steel reinforcement by changing the polarization of steel towards anodic and cathodic.

8.3 Inorganic Inhibitors

Inorganic inhibitors are inhibitors made from inorganic compounds, namely artificial chemicals from inorganic compounds. In other words, a chemical that is not carbon based. The most common inorganic inhibitors on the market are those made from nitride. The nitrides mixture is added to the concrete during casting. Recently, the use of this material has been reduced due to its effect on the strength of concrete and the possibility to be attacked by an alkali-silica reaction. Thus, it is very important to control the optimal amount of inhibitor added so that the inhibitor works properly [17]. Another example of an inorganic inhibitor is sodium mono-fluorophosphate (Na_2PO_3F). Sodium mono-fluorophosphate (Na_2PO_3F) has the ability to extend the life of concrete and reduce the corrosion rate. It can even be used on concrete that is attacked by carbonates. Quality of inhibitor is determined by ability to reduce corrosion rate. Table 8.1 shows performance of different types of inhibitor in reducing corrosion rate of steels in concrete exposed in chlorine solution.

8.4 Green Inhibitors

So far, inhibitors are made from inorganic chemicals and have been used extensively in the oil and gas and infrastructure industries. However, currently there is an effort to convert inorganic inhibitors into organic inhibitors as considering the environmental impact caused by inorganic inhibitors. Green inhibitors have now become a major concern in the selection of steel inhibitors for concrete. The advantage of this inhibitor is not only low cost, but also compatibility with concrete. In addition, green inhibitors are non-toxic and can be naturally degraded. Inhibitors work by changing the anodic or cathodic reaction which in turn affects the corrosion rate. Usually, the more inhibitors are used, the thicker the oxide film formed and the resistivity of the film increases.

Research on green inhibitors shows that green inhibitors are safe, biodegradable and environmentally friendly. Green inhibitors can also be found or easily produced. Table 8.2 presents types of plants that can be used for steel reinforcement inhibitors. Figure 8.4 is classification of green inhibitors used to reduce corrosion of reinforcing steel in concrete.

The following shows natural ingredients that can be used to make green inhibitors. The inhibitors of extracted black pepper and nicotine have been studied by Quraishi [21], and Quraishi [20]. Research result showed that this inhibitor can reduce the corrosion rate. They studied the corrosion inhibition of mild steel in chloride solution by extracts of black pepper (Piper nigrum family: *piperaceae*) by measurement

Table 8.2 Types of plants that can be used for steel reinforcement inhibitors [7]

Plant extracts	Element	Protection mechanism	Metal	Solution
Argemone Mexico	Alkaloids	Free electrons on the O and N atoms form bonds with electrons on the metal surface. The O atoms help free electrons on the N atoms and form stronger bonds with the electrons of the metal		
Garlic	Allyl propyl disulfide	Affects the potential cathodic process of steel		
Carrot	Ionized pyrrolidine	The N atom acquires a negative charge, and the free electrons in N have a still higher charge, resulting in the formation of a stronger bond in N		Alkali
Black pepper	Alkaloid “Piperine”	Mixed-type inhibitors and charge transfer control the corrosion process	Light steel	Hcl
Fennel Seeds	Limonene (20.8%) and Pinene (17.8%)	Adsorption on metal surface	Carbon steel	1 M HCl
Soybeans	Hydroxy aromatic compounds	Complex shape with metal		
Allium Cepa (Onion)	Quercetin	Adsorption sites on metal surfaces	Aluminum	Rubbish Water
Orange peel	Butyric acid, 2- bromo-, 1-methylethyl ester	Adsorption sites on metal surfaces	Light steel	Hcl

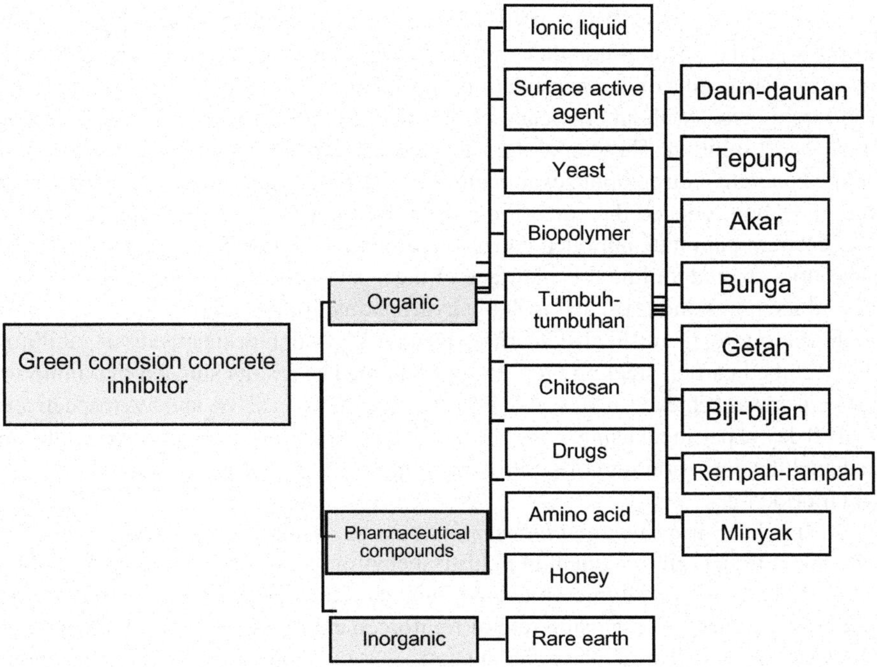

Fig. 8.4 Classification of green inhibitors used to reduce corrosion of reinforcing steel in concrete [4]

of mass loss, potentio-dynamic polarization, and electrochemical impedance spectroscopy (EIS). Black pepper extract gave maximum inhibition efficiency (98%) at 120 ppm and 35 °C for mild steel in hydrochloric acid medium. Corrosion inhibition properties are attributed to the "piperine" alkaloids. Nicotine is an organic alkaloid compound: a liquid, oily, colorless derivative of ortinine. This organic compound is a candidate for the protection of petroleum pipeline systems, because it comes from nature, easily found in the tobacco plant (nicotiana tabacum), where it is the main active chemical component.

8.4.1 Green Inhibitors for Steel Reinforcement

Inhibitors are able to form a protective layer that can prevent aggressive species such as oxygen, carbon-di-oxide, sulfates, chlorides, and moisture from coming into contact with the steel in the concrete. Thus corrosion does not occur. The effectiveness of inhibitors in preventing corrosion depends on the nature of the constituent materials. Many studies have been carried out on the use of inhibitors from plant extracts [6, 8, 9]. In the following research, it is presented organic materials that can be used as green inhibitors to protect steel reinforcement from corrosion.

Quraishi [20], studied the effect of calcium palmitate combined with calcium nitrite to make an inhibitor that can protect steel in concrete from the corrosion process. The results of the investigation showed that calcium palmitate could protect reinforcing steel with an efficiency of 91% to 92% after 90 days of testing time in 3.5% NaCl solution. These inhibitors also have no effect on the mechanical strength of the concrete. Petrographic examination revealed that calcium palmitate narrowed the pores and reduced the corrosion rate of the steel. Further investigations found that calcium palmitate inhibited corrosion through an adsorption mechanism. These inhibitors created a film on the surface of the steel via polar carboxylic groups that closed the pores and form an insoluble hydrophobic iron stearate salt.

Furthermore, Quraishi [19], learn performance of calcium stearate as an inhibitor. They embed carbon steel on OPC IS: 456–2000. The results showed that inhibitor efficiency was achieved at 90% and 93% at concentrations of 3% and 5% respectively in a 60 day experiment using 3.5% NaCl. Similar to previous studies, these inhibitors reduced the corrosion rate of steel by blocking porous concrete to limit the ingress of chloride ions.

Joshua [13], used *phyllanthus muellerianus* as an inhibitor to reduce corrosion of concrete steel reinforcement in industrial environments. They used 0.5 M H_2SO_4 to simulate an industrial/microbial environment. At a concentration of 6.67 g/l, this inhibitor reduced the corrosion rate of reinforcement by up to 90%. Meanwhile, at a concentration of 1.67%, the reducing corrosion rate was 78%. From the investigation, the leaves of *phyllanthus muellerianus* and *euphorbiaceae* contain constituents namely tannins, *phlatanins*, saponins, flavonoids, terpenoids and alkaloids.

Abbas [1], investigated the effectiveness of green inhibitors extracted from orange peel waste. They extracted dried orange peel using methanol extract at 6 h immersion time in methanol at pressure (60 mbar) and 40 °C. From experimental data using electrochemical polarization measurements and weight loss testing for 7 days of immersion time, the results showed that steel had decreased corrosion by 0.02 mm/year at 3% inhibitor concentration.

Shaymaa [23], studied the use of rice husks to prevent corrosion of steel in concrete. They added rice husk extract to the concrete using the American mix design method (ACI 211) [3]. Concrete reinforcement samples were immersed in water mixed with 3.5% NaCl solution for 30 days. Corrosion tests were carried out on solutions in different concentrations (1%, 2% and 3%) of inhibitors. The data noted that the corrosion current was 41.3 $\mu A/cm^2$ for the solution without inhibitor. Corrosion currents that occur are 28.5 $\mu A/cm^2$ and 7.8 $\mu A/cm^2$ after adding 1% and 3% rice husk inhibitors. This means a reduction in the corrosion rate is 30% and 81%, respectively.

Lotto [16], developed *vernonia amygdalina* (bitter leaf) as a corrosion inhibitor for reinforcement. Reinforcements with chemical composition: 0.3% C, 0.25% Si, 1.5% Mn, 0.04% P, 0.64% S, 0.25% Cu, 0.1% Cr, 0.11% Ni, and Fe, were used for reinforcement. Four different concentrations of 25%, 50%, 75% and 100% extract were used. Experiments were carried out using bitter leaf extract as a green inhibitor in 3.5% sodium chloride solution. Inhibitor extracts with concentrations of 25, 50, 75, and 100% were prepared from fresh leaves of *vernonia amygdalina* with distilled water. Vernonia amygdalina extract gave good corrosion inhibition performance to steel reinforcement in concrete at concentrations of 25%, 50% and 75% in NaCl test medium. The highest inhibition efficiency of 90.08% was achieved at a concentration of 25%.

An inhibitor extracted from the leaf form of bambusa arundinacea (Indian bamboo) for reinforcement protection has been investigated by *Abdul Rahman*, 2011 [2]. Ordinary portland cement (OPC) was used in this study. The concrete was mixed with a chloride concentration of 0.94%. The water content was kept constant up to 230 kg/m^3. The water-cement ratio (w/c) used was 0.45. From the experimental data, using the bambusa arundinacea extract decreased the reduction in the corrosion rate of the reinforcement to 1.53×10^{-3}. Compared without inhibitors, the corrosion rate of the reinforcement was 2.169×10^{-3} (EIS) and 1.8×10^{-3} (LPR) which means there was an efficiency of 72%. The LPR and EIS results also showed that the concrete resistivity (Rc) and polarization resistance (Rp) values increased with increasing concentrations of arundinacea bambusa.

Lisha [15], used powdered azadirachta indica (neem) and dehydrated aloe vera as corrosion inhibitors for steel in concrete. Class M 25 concrete was made using 20 mm coarse aggregate. The concrete was immersed in a solution with a salinity of about 3.5% (35 g/L). Concrete with green neem corrosion inhibitor had reduced the corrosion rate of steel in concrete from 0.3 mm/y to 0.22 mm/y. The aloe vera extract inhibitor showed that the corrosion rate was reduced to 0.27 mm/y. The inhibitor results showed that azadirachta indica (neem) had better corrosion inhibition efficiency compared to aloe vera inhibitors.

Akshatha [5], conducted research on inhibitors with ruta graveolens and azadirachta leaves. Extracted leaves of azadirachta indica (neem) and ruta graveolens plant were used as organic inhibitors. They compared with inorganic inhibitors namely sodium nitrate and ethylene diaminetetra acetate (EDTA) disodium dihydrate. Inhibitors were added during mixing of concrete and steel bars. The concrete was made of OPC 43 grade cement with a specific gravity of 3,279. The reinforcement was immersed in a solution of hydrochloric acid (HCl), sodium chloride (NaCl) and magnesium sulfate ($MgSO^4$) for corrosion testing. From the measurement of the half-cell potency, the experiment at 5% HCl for 56 days, it can be seen that the azadirachta Indica extract had shown the most positive potency followed by ruta graveolens, sodium nitrate and EDTA disodium dihydrate. The same trend also occurred when the experiment was carried out in 5% NaCl solution which showed that Azadirachta Indica provided the most positive corrosion potential. On the other hand, when the reinforcement was immersed in 5% $MgSO_4$ solution, the inorganic inhibitors (EDTA dehydrated disodium and sodium nitrate) showed the most positive corrosion potential.

Eyu [10], studied the application of vernonia amygdalina (bitter leaf) as a corrosion inhibitor on carbon steel reinforcement (class 40) immersed in 3.5% NaCl solution for 8 weeks of immersion time. They extracted the leaves using methanol to get liquid inhibitors. By using a cylindrical concrete block mixed with vernonia extract, they studied the corrosion rate of the reinforcement. In another experiment, they used sodium nitrite as an inorganic corrosion inhibitor for comparison. From observations vernonia extract contains alkaloids, saponins and tannins. The data showed that sodium nitrite at 2% v/v in 3.5% NaCl solution had a corrosion rate of 0.007 mm/y for 8 weeks of immersion time. The corrosion rate of the reinforcement without inhibitor was 0.09 mm/y. From these data they noted that the sodium nitrite inhibitor had the highest inhibition efficiency of 96%. The inhibitor vernonia amygdalina showed an increase in Ecorr to + 95 mV and calcium nitrite + 85 mV for a concentration of 12 l/m^3 in 70 days of immersion. Further investigating, they noted that it was the presence of tannins, alkaloids and saponnins in vernonia amygdalina that acted as a barrier for chloride entry. The weight loss test revealed that the sodium nitrite inhibitor showed a higher inhibition efficiency of 96% followed by calcium nitrite with 91% for the 2% v/v inhibitor. However, vernonia amygdalina had an inhibition efficiency of 75% with 6% v/v. The inhibitor vernonia amygdalina showed an increase in Ecorr to +95 mV and calcium nitrite +85 mV for a concentration of 12 l/m^3 in 70 days of immersion. Further investigation, they noted that it was the presence of tannins, alkaloids and saponnins in vernonia amygdalina that acted as a barrier for chloride entry.

Palanisami [18], investigated the effect of prosopis juliflora extract on steel corrosion in concrete in 3.5% NaCl. Corrosion tests were carried out after 30 days of immersion of the reinforcement without and with inhibitors. From the corrosion test it was revealed that the corrosion inhibitor would achieve 91% efficiency at an inhibitor concentration of 120 ppm. At a low inhibitor concentration (100 ppm), the efficiency was 51%. From the EIS study, it was stated that there was a diffusion process and a precipitation effect of the solid calcium hydroxide layer for the formation of a protective layer at the steel/concrete interface. Furthermore, increasing the inhibitor concentration will reduce the surface inhomogeneity due to the adsorption of extract molecules on the surface of the steel reinforcement. From the AFM image, it can be seen that there was absorption of inhibitor molecules onto the steel surface to form a protective layer which reduced the corrosion rate of the reinforcement.

8.5 Inhibitor Testing

Inhibitor testing is carried out to ensure the success of the inhibitor in reducing the corrosion process. Testing is useful to know the behavior of the inhibitor if it will be used in various working conditions and different materials. Table 8.3 guide number of inhibitor should be applied to reach protection system for the steel in concrete. Each inhibitor has advantages and disadvantages and reacts differently under different working conditions. Thus, by carrying out a series of tests, an inhibitor can be selected according to the material and conditions to be used. Selection is based on efficiency, availability, cost, ease of application and environmental impact criteria. Experimental data is used to determine how much needs, transportation costs, storage, and distribution costs.

The general criteria for choosing an inhibitor are done by testing the inhibitor. The test aims to observe the behavior of inhibitors in terms of: ability to reduce corrosion,

Table 8.3 Criteria in determining the amount of inhibitor to be used [24]

Corrosion severity	The amount of inhibitor to be used (ppm)
Currently	10–15
Intermediate	15–25
High	>25

consistency in reducing corrosion, ability to adapt to changing environments, speed distribution over the entire surface of the reinforcing steel, effect on concrete, ability to form a layer on the reinforcing steel, quality of the film layer formed, and the tendency of evaporation, and ability for thickening.

8.5.1 Inhibitor Performance

The performance of the inhibitor is indicated by the effectiveness which is the percentage of the reduction in the corrosion rate after the inhibitor is added. Corrosion inhibitors reduce the rate of corrosion by forming a film that inhibits the corrosion process on the metal surface. Another mechanism is to increase the resistivity of the solution to reduce the corrosion rate. Inhibitor efficiency is expressed by the formula:

$$\frac{[CR\text{uninhibited} - CR\text{inhibited}]}{CR\text{inhibited}} \times 100\% \tag{8.1}$$

or

$$\frac{[Rp, \text{inhibited} - Rp, \text{uninhibited}]}{Rp, \text{inhibited}} \times 100\% \tag{8.2}$$

where CR is the corrosion speed and Rp is the polarization resistance.

The quality of the inhibitor determines the ability of the inhibitor to inhibit the corrosion rate. Some conditions that must be considered in order to prevent failure of the inhibitor are: formation of foam, emulsion, blockage, thinning of the film, safety, and handling. Foam can occur in places where there is turbulence flow which results in the mixing of fluid and air. Table 8.4 measures corrosion rate caused by different method of injection in the concrete. Compatibility of inhibitors and metals must also be considered. For example amine based inhibitors protect steel but attack copper and brass surfaces. Reduction of nitrate inhibitors produces ammonia (NH_3) which causes stress corrosion cracking (SCC) of metals made of copper and brass. Hence, it is important to know the metal component of a system to ensure compatibility of the inhibitor with each metal to be protected.

Table 8.4 Effect of injection system on the performance of reinforcing steel inhibitors in concrete [12, 14]

System	Cement	E_{corr} (mV vs. SCE)	I_{corr} (mA.cm^{-2}) × 10^{-5}	Corrosion rate (mmpy) × 10^{-3}	Inhibitors efficiency (%)
Without electro injection	OPC	−591	91,910	10,650	–
	PPC	−567	82,080	9,512	–
	PSC	−478	30,690	3,556	–
With electro injection	OPC	−282	6034	0.699	93.43
	PPC	−345	2,276	0.263	97.22
	PSC	−295	2024	0.234	93.40

References

1. Abbas, A.S., Fazakas, É., Török, T.I.: Corrosion studies of steel rebar samples in neutral sodium chloride solution also in the presence of a bio-based (green) inhibitor. Int. J. Corrosion Scale Inhib. **7**(1), 38–47 (2018)
2. Abdulrahman, A.S., Ismail, M., Hussain, M.S.: Inhibition of corrosion of mild steel in hydrochloric acid by Bambusa arundinacea. Int. Rev. Mech. Eng. **5**(1), 59–63 (2011)
3. ACI 211.2–98: ACI Standard Practice for Selecting Proportions for Structural Lightweight Concrete (2004)
4. Ahmed, E.S., Ganesh, G.M.: A Comprehensive overview on corrosion in RCC and its prevention using various green corrosion inhibitors. Buildings **12**(10), 1682 (2022)
5. Akshatha, G.B., Kumar, P.: Effect of corrosion inhibitors in reinforced concrete. Int. J. Innov. Res. Sci. Eng. Technol. **4**(8) (2015)
6. Asmara, Y.P.: Long term corrosion experiment of steel rebar in fly ash-based geopolymer concrete in Nacl solution, Hindawi. Int. J. Corrosion **1**(1) (2016)
7. Asmara, Y.P.: Application of plants extracts as green corrosion inhibitors for steel in concrete—a review. Indonesian J. Sci. Technol. **3**(2) (2018)
8. Asmara, Y.P.: Corrosion Inhibitor Laboratory Report. University Technology Petronas, Malaysia (2019)
9. Asmara, Y.P.: Development of green vapor corrosion inhibitor. IOP Conference Series: Materials Science and Engineering 257, 012089 (2017)
10. Eyu, D.H.E., Chukwuekezie, J., Idris, M.: Effect of green inhibitor on the corrosion behavior of reinforced carbon steel in concrete. ARPN J. Eng. Appl. Sci. 8(5) (2013)
11. Fontana, M.G.: Corrosion Engineering, 3rd edn. McGraw-Hill, New York (1986)
12. Han-Seung, L., Saraswathy, V., Kwon, S.-J., Karthick, S.: Corrosion Inhibitors for Reinforced Concrete: A Review, Books Corrosion Inhibitors, Principles and Recent Applications, Intech (2017)
13. Joshua, O., Olubanke, O.: Investigating prospects of Phyllanthus muellerianus as ecofriendly/sustainable material for reducing concrete steel reinforcement corrosion in industrial/microbial environment. Energy Procedia **74**, 1274–1281 (2015)
14. Karthick, S.P., Madhavamayandi, A., Murlidharan, S., Saraswathy, V.: Electrochemical process to improve the durability of concrete structures. J. Build. Eng. **7**, 273–280 (2016)
15. Lisha, C., Sunilaa, G.: Corrosion resistance of reinforced concrete with green corrosion. Int. J. Eng. Sci. Invention Res. Dev. **3**(11) (2017)
16. Loto, C.A., Loto, P.: Inhibition effect of vernonia amygdalina extract on the corrosion of mild steel reinforcement in concrete in 3.5 M NaCl environment. Int. J. Electrochem. Sci. **8**, 11087–11100 (2013)

17. Luna, M., Rivetti, S.: Corrosion Inhibitors for Reinforced Concrete. Intechopen, 72772 (2017). https://www.intechopen.com/chapters/58498
18. Palanisamy, S.P., Kamal, Venkatesh, G.: Prosopis juliflora—a green corrosion inhibitor for reinforced steel in concrete. Res. Chem. Intermediates **42**(12), 7823–7840 (2016)
19. Quraishi, M.A., Abhilash, Singh, B.N.: Calcium stearate: a green corrosion inhibitor for steel in concrete environment. J. Mater. Environ. Sci. **2**(4), 365–372 (2011), ISSN: 2028-2508
20. Quraishi, M., Singh, B.N., Singh, S.K.: Calcium palmitate: a green corrosion inhibitor for steel in concrete environments. J. Mater. Environ. Sci. X(6), 1001–1008, 2028–2508 (2012)
21. Quraishib: Green approach to corrosion inhibition of Mild steel in Hydrochloric acid and sulfuric acid solutions by the extract of Murraya koenigii leaves. Mater. Chem. Phys. (122), 114–122 (2011)
22. Saraswathy, V., Song, H.W.: Improving the durability of concrete by using inhibitors. building and environment. Build. Environ. **42**(1), 464–472 (2007). https://doi.org/10.1016/j.buildenv.2005.08.003
23. Shaymaa, A., Ali, I., Al-Mosawi, Amer, A., Hadi: Studying the effect of eco-addition inhibitors on corrosion resistance of reinforced concrete. Bioprocess Eng. **1**(3), 81–86 (2017)
24. Viswanathan, S., Savior, S., Umoren: Corrosion Inhibitors in the Oil and Gas Industry, Wiley-VCH. Issue 1 (2020)

Chapter 9
Geopolymer Concrete

Abstract This chapter introduces the concept of geopolymer concrete, an innovative and remarkable alternative material known for its exceptional resistance against corrosion. The primary focus of this exploration is to evaluate the ability of geopolymer concrete to withstand corrosive environments, while also conducting a comprehensive examination of the various types of resins employed for its protection. Throughout the chapter, there is an in-depth discussion on the composition of geopolymer materials, which includes notable elements such as fly ash (FA), palm oil fuel ash (POFA), kaolin, metakaolin, and dolomite. These materials play a pivotal role in the geopolymerization process, and their chemical composition and interactions are thoroughly explored and analyzed. By explaining the complexities of geopolymer concrete, this chapter provides valuable insights into its remarkable corrosion resistance. Furthermore, the chapter emphasizes the significance of activators and their profound influence on the properties of geopolymer materials. Additionally, it explores a range of factors that impact the properties of geopolymer concrete, including the selection of raw materials, the precise conditions governing the curing process, the choice of aggregates, and the incorporation of additives, nanomaterials, and fibers. Moreover, the chapter highlights various applications of geopolymer concrete. Overall, this chapter provides a comprehensive understanding of geopolymer concrete, examining the role of activators, analyzing influential factors, and exploring its diverse range of applications in construction and material engineering.

9.1 Background

One effort to improve the quality of concrete is to make geopolymer concrete. Geopolymer concrete made from inorganic polymer materials has the potential to replace the use of OPC as a building material. Geopolymer can be made from coal industry waste material and smelting furnace slag where the material contains aluminosilicate. Geopolymer concrete uses materials other than cement as adhesives. As a substitute for cement, an adhesive material containing the elements alumina and silacat is used, which is abundant in fly ash, bottom ash, and risk hush ash. The

Y. P. Asmara, *Concrete Reinforcement Degradation and Rehabilitation*, Engineering Materials, https://doi.org/10.1007/978-981-99-5933-4_9

use of geopolymer is carried out as an effort to prevent corrosion of reinforcing steel. The corrosion behavior of reinforcing steel in geopolymer concrete made of fly ash material shows that the reinforcement has increased corrosion resistance. Compared to OPC concrete, the use of geopolymer concrete can reduce the corrosion current density by 10% [1]. The same experiment showed that the geopolymer concrete cracked after 9 days while the OPC concrete was 2.5 days during the electrochemical experiment. Geopolymer concrete has nano-sized porosity so that it can withstand ion diffusion currents from outside. In addition, silicate membranes are able to form a strong layer on geopolymer reinforcement [2]. A study shows that geopolymer concrete is able to increase flexure strength by up to 100% and reduce the possibility of cracks by 24% [3]. Recently, the use of polymers to improve the quality of concrete has attracted attention and has begun to receive great attention. There are various advantages applying geopolymer concrete for construction as follows [4, 5]:

- Produces high strength and can be made with a size that is thin and light.
- Longer service life.
- Good creep strength, low shrinkage coefficient, sulfate resistance and corrosion resistance.
- Impermeable to salt attack.
- Produce a flexible shape suitable for use in architectural materials.
- Reducing environmental pollution due to CO_2 emissions.
- Environmentally friendly,
- Can recycle combustion and agricultural waste.
- Low maintenance costs.

However, geopolymer concrete has weaknesses compared to OPC concrete, especially in terms of:

- Price. Although the price of fly ash is cheap, the alkaline solution is relatively expensive.
- Strength. Geopolymer concrete is brittle, so cracks will easily spread to other places. But this can be overcome by applying fiber.
- Curing method. Geopolymer requires a high curing temperature to accelerate drying in order to obtain concrete with high strength properties.
- Steel-cement bond strength. The bond between reinforcing steel and concrete has not been studied as much as OPC.

9.2 Geopolymers Manufacturing

Geopolymer material is made through a geopolymerization reaction in which a process of changing the aluminosilicate raw material into a network consisting of $[-Si-O-Al-O-]_n$ bonds $[-Si-O-Al-O-]_n$ occurs [6]. In other words, a geosynthesis reaction occurs, namely the chemically integrated synthesis of minerals (see Fig. 9.1). The geopolymerization reaction results in the formation of a binder which can harden to form a geopolymer material. Geopolymers can be made by utilizing

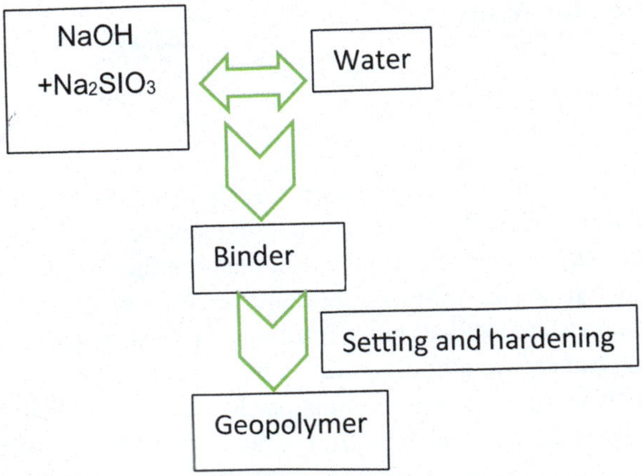

Fig. 9.1 Schematic of the geopolymer manufacturing process [7]

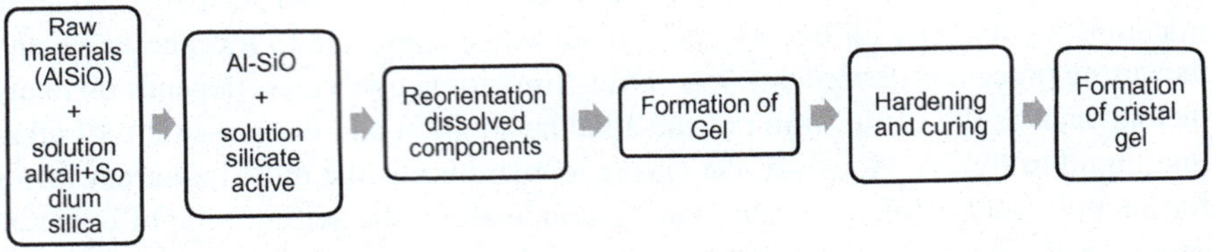

Fig. 9.2 The process of forming geopolymer concrete

different aluminosilicate sources in nature such as red mud, blast furnace slag, kaoli nite, and rice husk ash. Such materials play an important role in determining t physicochemical and mechanical properties of geopolymer materials (see Fig. 9

Conventionally, geopolymer is made by mixing amorphous aluminosili fly ash with sodium hydroxide and sodium silicate solutions which can fo geopolymer gel. The stages of geopolymer formation are: (a) association; molecules react with siloxane bonds (–Si–O–Si–) found in aluminosilicate ra rials. (b) Dissociation; Intermediate pentavalent silicon dissociates in an i manner and forms silanol (>Si–OH) and aluminol (>Al–OH) groups.

The geopolymerization process is a chemical reaction between an all tion and a source material containing aluminosilicate (FA) and produ dimensional polymer chain and a ring structure consisting of Si–O–Al reported by Sitaram [6]. The reaction can occur at room temperature. H considered as an energy efficient and much cleaner source. Si and A in fly ash can be dissolved by the action of hydroxide ions. Precu converted into monomers. Polycondensation of monomers into pol

9.3 Geopolymer Materials

9.3.1 Fly Ash (FA)

FA is a residue from burning coal which is widely available throughout the world and causes pollution. Thus, fly ash-based geopolymer concrete is a good alternative to overcome the abundance of fly ash. FA is a material that is often used in making geopolymer concrete for the following reasons: resistant to attack by sulfuric acid, small shrinkage, resistant to alkali-silica reactions, fire resistant, does not pollute the environment and can reduce air pollution.

Fly ash is rich in silicates and alumina, so it reacts with alkaline solutions to produce aluminosilicate gels which bind aggregates to produce good concrete. In fly ash-based geopolymer concrete, the silica and alumina present in the source material are first induced by an alkaline activator to form a gel known as an aluminosilicate. The materials commonly used as alkaline solutions in the manufacture of fly ash-based geopolymers are sodium silicate and potassium hydroxide [8]. Then these aterials are mixed with fine aggregate and coarse aggregate to form concrete and curing process is carried out. The setting time for geopolymers depends on many rs such as the composition of the alkaline solution and the mass ratio of alkaquid to fly ash. However, the curing temperature is the most important factor polymers. With increasing curing temperature, the setting time of concrete [9]. During the curing process, geopolymer concrete undergoes a polymerocess. Due to the increase in temperature, polymerization becomes faster te can gain 70% of its strength within 3 to 4 h of curing [10]. Figure 9.3 s of alkali-activated fly ash used in geopolymer concrete.

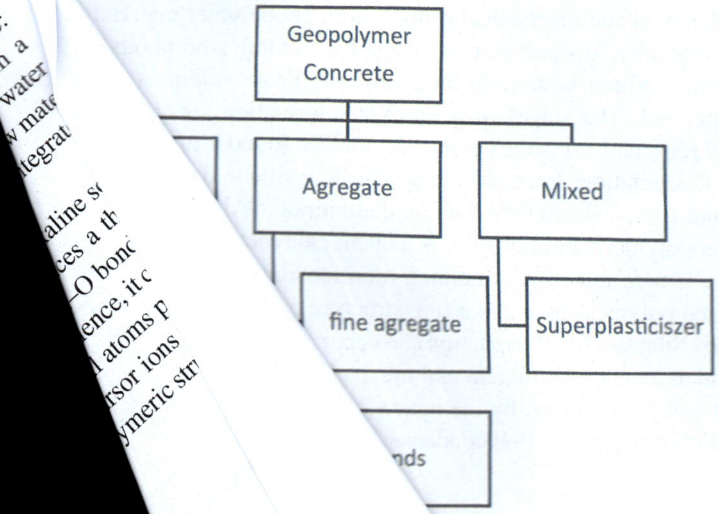

crete [4]

9.3.2 *Palm Oil Fuel Ash Ash (POFA)*

POVA comes from palm oil waste such as fibers, shells, palm kernel and bunches which are solid wastes that can be obtained from palm oil mills. To produce POVA, the waste is burned in a boiler to produce two types of palm ash, namely boiler ash and palm oil fuel ash (POFA).

9.3.3 *Kaolin*

Kaolin contains a lot of aluminosilicate which is fine clay, kaolin is commonly used to make ceramics. The aluminosilicate sources that are often used are kaolinite, fly ash, calcined kaolin, and chemically synthesized kaolin. Geopolymers are synthesized by polycondensation under 100 °C at room pressure in alkaline solution. Kaolin is soft, non-abrasive, insulating, high electrical resistance. Thus, it can be concluded that the secondary minerals contained in kaolin will affect the reaction process and geopolymer properties (Zain 2016).

9.3.4 *Metakaolin*

Metakaolin is the dehydroxylated form of the mineral kaolinite. Metakaolin is obtained by calcination or dehydroxylation of kaolin clay at 500–900 °C. This method can remove chemically bound water and change most of the aluminum octahedral coordination (Zain 2016).

9.3.5 *Dolomites*

Dolomite is relatively soft and easily crushed into a fine powder. Dolomite can be u
to reduce acidity. Dolomite is a rock-forming mineral containing calcium, magnes
carbonate $CaMg(CO_3)_2$. Dolomite is a major component of the sedimentary
known as dolostone and the metamorphic rock known as dolomite marble. Lime
containing some dolomite is known as dolomitic limestone. Table 9.1 shows che
composition of various geopolymer materials. Table 9.2 presents elements u
form geopolymer materials.

Table 9.1 Chemical composition of various geopolymer materials using XRF [11]

Chemical composition	Fly ash	POFA	Kaolin	Metakaolin	Dolomites
SiO_2	52.11	51.18	52.00	55.90	15.37
Al_2O_3	23.59	4.61	35.00	37.20	1.69
Fe_2O_3	7.39	3.42	1.00	1.70	0.51
TiO_2	0.88		0.90	2.40	0.015
CaO	2.61	6.93	<0.05	0.11	23.00
MgO	0.78	4.02	0.70	0.24	17.20
K_2O	0.80	5.52	2.00	0.18	0.195
Na_2O	0.42	0.06	0.05	0.27	0.013
SO_3	0.49			0.02	
P_2O_5	1.31			0.17	0.019
Loss in ignition		21.6		0.80	

Table 9.2 The highest chemical composition of geopolymer materials [11]

Various types of geopolymer based	Highest chemical composition	Sources of raw materials
'y ash	SiO_2 & Al_2O_3	Waste product from coal mining
·kaolin	SiO_2 & Al_2O_3	Natural resources
	SiO_2 & Al_2O_3	Natural resources
	SiO_2	Waste product from palm oil industry
	CaO, MgO & SiO_2	Natural resources

vator

ıterials that need to be mixed into the geopolymer cement mixture
ɔccurs between the concrete components. There are two types of
activator and acidic activator. Alkaline activator is a type of acti-
ın alkaline environment (high pH) and accelerate the hydration
alkaline activators are: SiO_3 and NaOH. lithium hydroxide,
f activator is acidic activators, for example: phosphoric acid.

sed
um
ock
stone
mical
sed to

9.4 Factors Affecting the Properties of Geopolymers

9.4.1 *Raw Material*

The most frequently used raw material is fly ash (FA) particles. FA determines the geopolymer properties, the finer the FA particles, the greater the surface area of FA to form bonds with the constituent materials and cause high durability.

9.4.2 *Activators*

The type of alkaline activator affects the mechanical properties and microstructure of the geopolymer. For different adhesives, the 7-day compressive strength of the geopolymer decreased as the ratio of alkaline solution to binder increased from 0.35 to 0.50.

9.4.3 *Curing Conditions*

The geopolymer strength varies with the curing time. Geopolymer mortar made with Na_2SiO_3 and NaOH solution and cured at room temperature has high compressive strength [12].

9.4.4 *Aggregate*

The geopolymer strength decreases with aggregate content. The use of fine dolomite aggregate increases the mechanical strength of concrete by 175%, compared to using sand which increases 78% [13].

9.4.5 *Additives, Nano Materials and Fibers*

The addition of silica fume (SF) to geopolymer concrete can increase mechanical strength, but excessive SF is harmful to geopolymers [14]. The compressive strength of the geopolymer mortar containing 4% nano-meta kaolin increased significantly [15]. It was found that the addition of silica micro powder increased geopolymer activity, but decreased FA activity [16].

9.5 Application of Geopolymer Concrete

9.5.1 Repairing Concrete

The tensile and adhesive strength of geopolymer concrete at 28 days is similar to that of commercial concrete filling materials and has the potential to be used as a concrete filling material [17]. Many studies have studied that the addition of 14 M sodium hydroxide solution and 10% calcium carbonate can increase the shear bond and flexural strength of geopolymer mortar [18].

9.5.2 Construction of Buildings on the Sea

Reinforced concrete exposed to seawater, rainwater and soil containing kaolin will cause damage over a long time. The potential of geopolymer concrete which is resistant to chemical attack and especially sulfate resistance makes geopolymers have hope of being used for marine construction materials. Compared to OPC concrete, geopolymer concrete contains more amorphous phases, smaller porosity and density which makes it difficult for seawater to seep.

9.5.3 Road Pavement Materials

Several studies have proven that geopolymer materials such as FA and slag can be used as basic materials for road pavements. This not only contributes to the sustainable use of waste by-products, but can also reduce the use of cement to a certain extent, which is a low-carbon approach.

9.5.4 Fire and High Temperature Resistant Material

Geopolymerization is a material for converting industrial solid waste into a chemically resistant cement binder with excellent thermal stability and is difficult to burn.

9.5.5 Insulation Material

Geoplymer material has low electrical and thermal resistance so it can be used as a thermal insulation material. Geopolymers can also be used for soundproofing rooms.

Table 9.3 Thermal and acoustic properties of partial geopolymer materials [19]

Material	Characterization
Geopolymer pastes	Energy-efficient (15%)
Geopolymers and phase change materials	Energy-efficient (28%e30%)
Vegetal geopolymer concrete	Energy-efficient (46%e58%)
FA-based geopolymeric foams	Sound absorption coefficient (0.23 at 500 Hz, 0.25 at 2000 Hz)
Kaolinite-based porous geopolymer	Sound absorption coefficient (0.539 at 5000 Hz, 0.809 at 500 Hz)

9.5.6 Absorbent Material

As a cleaner production material, geopolymer can be used as an adsorbent to remove metal ions from wastewater (Table 9.3).

9.6 Hydrocarbon Adhesives

To improve the quality of geopolymer concrete, aggregate and cement replacement materials are often added. The most common use of cement replacement materials are hydrocarbon-based adhesives, namely epoxy and various types of resins. Resin-based materials can be used to bond objects to concrete or to bond concrete together. Each type of concrete adhesive has different strengths and weaknesses, and each is suitable for use in different situations. Resin and epoxy based materials have been widely used for the construction industry. This type of material is usually used as a surface coating material or for structural repair and decoration. For broader purposes, the use of epoxy and resin is still rarely used considering the price is still relatively expensive. Besides that. Resin is a liquid organic substance that can harden and is not soluble in water. Resins are a basic component of plastics and are also used to glue paper, particleboard, and other surfaces that require a decorative, protective, or special purpose finish.

9.6.1 Types of Resins

Epoxy is a type of adhesive that is very hard and weather resistant. Epoxy can be used for cement replacement adhesives and can be used for long term purposes, is durable, able to withstand extreme weather, withstands temperature, exposure to UV light, and even certain types of chemical exposure cement may be exposed to. Epoxy is often used to bond concrete or cement in basements or in exterior applications, and certain epoxy modified mortars can be used to bond concrete blocks or other building materials together. However, epoxy is still expensive, and its strength cannot be used for all types of construction. Epoxy has the added advantage of being resistant to shrinkage and wear and tear once it has dried.

a. ***Melamine–formaldehyde (MF) resin***

This resin made from melamine and formaldehyde (MF) is a thermoset MF which is more resistant. MF resins are used to produce plywood and particleboard for exterior or semi-exterior applications. Melamine is a very expensive chemical, so MF resins are usually made with urea (MUF resin) to reduce costs [20].

b. ***Resin polyester concrete (PRC)***

PRC is a thermosetting polyester resin that can be mixed into concrete. PRC behaves like plastic.

c. ***Acrylic resin***

Acrylic resin is made from acrylic acid, methacrylic acid. They are clear with good tensile strength. Impact and UV resistant acrylic resin. They are used for structural and decorative panels, adhesives, elastomers, coatings, signs, and translucent tiles. Polystyrene Resin. Polystyrene resin is an aromatic hydrocarbon polymer made from styrene monomer. The resin has excellent resistance to acids, alkalis and salts. Polystyrene resin has good clarity and flexibility. Used for insulation, pipes, foam, cooling towers, rubber, automotive instruments.

d. ***Silicone Resin***

Silicone resin is made with sodium silicate and various chlorosilicates. It can also be prepared with the less reactive tetraethoxysilane or ethyl polysilicate and some disiloxanes. This resin has good thermal and oxidative stability and is flexible and water repellent. Silicone resin forms a hard film due to its three-dimensional network structure. They are used for rubber, laminates, encapsulating resins, defoaming, and water-resistant applications.

9.7 Corrosion of Geopolymer Concrete

An aggressive environment causes OPC concrete to suffer damage due to corrosion. Damage can occur at the beginning of the manufacture of concrete structures due to the use of water containing chloride. The presence of chloride interferes with the passive condition of the concrete because of the alkalinity of the concrete which protects the steel surface. However, the effects of carbon, chlorine and acid conditions can damage the passive film which corrodes reinforcing steel exposed to the active environment. Recently, the use of polymers to improve the quality of concrete has attracted great attention and has begun to receive great attention. Geopolymer concrete is proven to reduce concrete pores so that it can resist the entry of chemical ions into the reinforcing steel [21, 22].

Eventhough polymer concrete is watertight, the presence of liquid cannot be avoided. Water is used as a solvent for aggregates and resins. Resins are also chemical compounds that may affect the corrosion properties of reinforcing steel. The use of fly ash as polymer concrete shows positive properties on corrosion speed. Polymer concrete using fly ash of 10, 30 and 50% has reduced the corrosion rate of steel in geopolymer concrete [23].

Figure 9.4 shows the polarization diagram of geoplymer concrete using fly ash of 0.10, 30 and 50%, the polarization diagram is measured after the concrete is 10 weeks old. Concrete A is concrete using OPC, concrete B, C, and D using OPC cement and fly ash of 10, 30 and 50%. Concrete E is a geopolymer in which fly ash is mixed with an alkaline solution, NaOH. Concrete B shows the smallest absolute current value and concrete E is the highest.

Based on Fig. 9.5 and Table 9.4, it can be seen that increasing the percentage of fly ash substitution in concrete to a certain amount will reduce the corrosion potential of concrete. From the experiments, concrete C (70% OPC and 30% fly ash) has the lowest corrosion potential. For concrete D (50% OPC and 50% fly ash), the corrosion potential is lower than concrete C. Concrete E (geopolymer concrete) has the highest negative corrosion potential even though it contains high amounts of fly ash. The reason may be that fly ash contains high amounts of metal oxides. If there is too much metal oxide contamination, the electrical conductivity will be high. High conductivity will allow the transfer of reinforcing metal ions to the concrete because there is a potential difference. Substitution of fly ash increases the resistance of concrete to corrosion. Increasing the percentage of fly ash substitution decreases the

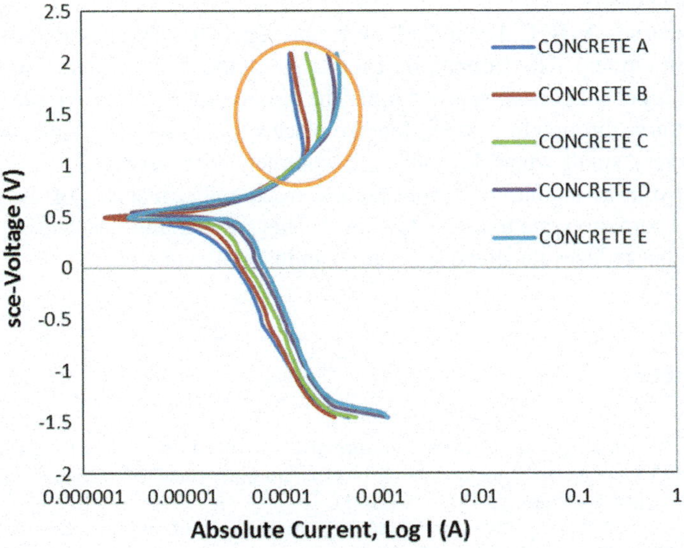

Fig. 9.4 Polarization diagram of various concrete conditions. Concrete A is concrete using OPC, concrete B, C, and D using OPC cement and fly ash of 10, 30 and 50% respectively. Concrete E is a geopolymer in which fly ash is mixed with an alkaline solution, NaOH [23]

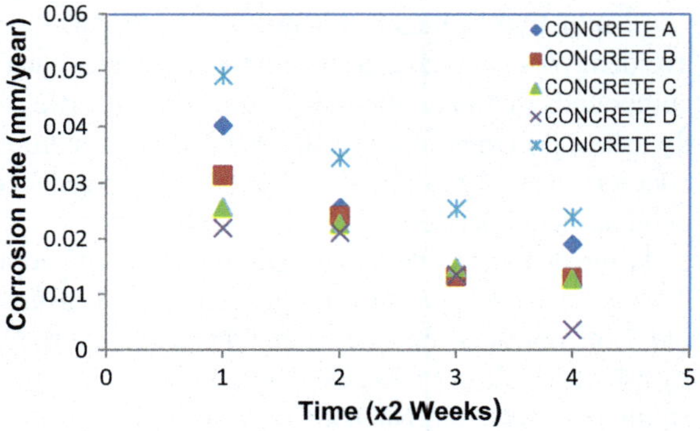

Fig. 9.5 Effect of the addition of fly ash on the corrosion rate of geopolymer concrete. Concrete A is concrete using OPC, concrete B, C, and D using OPC cement and fly ash of 10, 30 and 50%. Concrete E is a geopolymer in which fly ash is mixed with an alkaline solution, NaOH

Table 9.4 The effect of using fly ash on the corrosion rate of steel reinforcement [23]

Composition fly ash (%)	Corrosion potential, (Ecorr) V	Corrosion current, (Icorr) $\mu A/cm^2$	Corrosion rate, (CR) μm/yr
10	−0.549	1.1258	0.0131
30	−0.574	1.2208	0.0142
50	−0.585	1.0908	0.0126

corrosion potential of the reinforcement. This may be due to the smaller size of the fly ash particles than the cement so that they fill the pores formed in the concrete. As a result, the porosity of the concrete will decrease and chloride ions cannot penetrate the concrete easily.

All concretes A, B, C, D and E show the presence of stable passive films formed (marked in circles). The formation of a passive oxide film on metal surfaces is an important aspect of corrosion protection. The passive layer makes a dense passage so that metal ions through the oxide film take place very slowly. The ionization current becomes very small when the surface is completely covered with the oxide film. The passive layer is able to prevent the electrochemical reaction of the metal and reduce the corrosion rate to a very low level. This passive film does not actually stop corrosion but reduces the corrosion rate to an insignificant level.

References

1. Olivia, M., Nikraz, H.R.: Corrosion performance of embedded steel in fly ash geopolymer concrete by impressed voltage method. Incorporating Sustainable Practice in Mechanics of Structures and Materials, pp 781–786 (2011)
2. Gourley, J.T., Johnson, G.B.: Developments in geopolymer precast concrete article. J. World Congr. Geopolymer 139–143 (2005)
3. Sobhan, K., Martinez, F.J., Reddy, D.V.: Corrosion resistance of fiber-reinforced geopolymer structural concrete in a simulated marine environment, Can. J. Civil Eng. (2021)

4. Zerfu, K., Ekaputri, J.J.: Review on alkali-activated fly ash based geopolymer concrete. Mater. Sci. Forum 841, 162–169 (2016)
5. Gholampour, A., Ho, V.D., Ozbakkaloglu, T.: Ambient-cured geopolymer mortars prepared with waste-based sands: mechanical and durability-related properties and microstructure. Compos. Part B Eng. **160**, 519–534 (2019)
6. Sitaram, S.: Advanced geopolymerization technology. Books, Geopolymers and Other Geosynthetics (2019)
7. Singh, N.B.: Fly ash-based geopolymer binder: a future construction material. Minerals **8**(7), 299 (2018). https://doi.org/10.3390/min8070299
8. Abdullah, M.M.A., Kamarudin, H., Mohammed, H., Khairul Nizar, I., Rafiza, Zarina, J.: Review on fly ash-based geopolymer concrete without Portland Cement. Eng. Technol. Res. **3** (2011)
9. Chanh, N., Trung, A.B., Tuan, D.V.: Research geopolymer concrete. The 3rd ACF International Conference-ACF/VCA, 1–5 (2008)
10. Kong, D.L., Sanjayan, J.G.: Cemhical concrete. Composites **30**, 986 (2008)
11. Zain, H., Abdullah, M.M.A.B.: Review on various types of geopolymer materials with the environmental impact assessment. MATEC Web of Conferences (2017)
12. Huang, G., Yongsheng, J.L., Hou, Z.: Improving strength of calcined coal gangue geopolymer mortars via increasing calcium content. Constr. Build. Mater. **166**, 760–768 (2018)
13. Cong, P., Cheng, Y.: Advances in geopolymer materials: a comprehensive review. J. Traffic Transp. Eng. **8**(3), 283–314 (2021)
14. Cheah, C.B., Tan, L.E., Ramli, M.: The engineering properties and microstructure of sodium carbonate activated fly ash/slag blended mortars with silica fume. Compos B Eng **160** (2019)
15. Kaur, M., Singh, J., Kaur, M.: Microstructure and strength development of fly ash-based geopolymer mortar: role of nano-metakaolin. Constr Build Mater **190** (2018a)
16. Dong, M.H., Elchalakani, M., Karrech, A.: Development of high strength one-part geopolymer mortar using sodium metasilicate. Constr Build Mater **236** (2020)
17. Yodsudjai, W., Suwanvitaya, P., Pikulprayong, W.: (2010). Testing of geopolymer mortar properties for use as a repair material. Design Dev. Appl. Eng. Ceram. Compos. **215**, 325–334 (2010)
18. Phoo-ngernkham, T., Sata V., Hanjitsuwan, S.: High calcium fly ash geopolymer mortar containing Portland cement for use as repair material. Const. Build. Mater. **98**, 482–488 (2015)
19. Cong, P., Cheng, Y.: Advances in geopolymer materials: a comprehensive review. J Traffic Transp Eng (English Edition), **8**(3), 283–314 (June 2021)
20. Frazier, C.E.: Solid Wood Processing, Adhesions and Adhesives, Encyclopedia of Forest Sciences, pp. 1261–1269 (2004)
21. Fontana, M.G.: Corrosion Engineering, 3rd edn. McGraw-Hill, New York (1986)
22. Uhlig, Revie, R.W.: Corrosion and corrosion control: an introduction to corrosion science and engineering, 3rd edn. Wiley, New York (1985)
23. Asmara, Y.P.: Long term corrosion experiment of steel rebar in fly ash-based geopolymer concrete in nacl solution, Hindawi. Int. J. Corros. 1(1)

Chapter 10
Concrete Treatment

Abstract This chapter explores various aspects of concrete treatment, focusing on re-alkalization, chloride removal, curing, coating, additives, and rust cleaning. The principle of re-alkalization in concrete is discussed, highlighting its significance in restoring alkalinity and protecting against corrosion. The process of dichlorination, involving the removal of chlorides, is also examined as an essential step in concrete treatment. Additionally, the importance of curing and coating in enhancing concrete durability and performance is explored, along with an explanation of the functions and types of coatings used. The use of additives to improve concrete properties is addressed, along with techniques for rust cleaning and guidelines on how to effectively clean rust from concrete surfaces. This chapter presents a valuable information regarding the approaches and treatments employed to improve the quality and lifespan of concrete structures.

10.1 Concrete Treatment

Reinforced concrete is a sturdy structure that can withstand all weather conditions. The combination of concrete and steel makes reinforced concrete have high compressive and tensile strength. However, reinforced concrete can be damaged if the quality of the concrete does not meet the requiremets standards. Over time, the strength of concrete can decrease. The strength of concrete will change either due to changes in the properties of the reinforcing steel or the concrete. The service life of concrete can be very long and very short depending on the loads applied and the chemical reaction that attack the concrete. The severity of the chemical attack on concrete depends on the location where the concrete is located. If the concrete is in a seawater environment, the concrete will experience attack by chloride elements, abrasion, vibration, and wind. The more complex the attack received by the concrete, the faster the concrete will damage. Techniques to extend the life of concrete can be done in two ways. It can be conducted before the concrete is fabricated or during in service. In selecting a reinforced concrete treatment method, various considerations need to be evaluated. Certain types of concrete have different reactions in different environmental conditions.

Y. P. Asmara, *Concrete Reinforcement Degradation and Rehabilitation*, Engineering Materials, https://doi.org/10.1007/978-981-99-5933-4_10

The resistance of concrete to damage is influenced by several factors:

- Thermal properties of concrete and steel materials. Each material has a coefficient of expansion and shrinkage which will have an impact on the concrete-steel bond strength.
- Electrical properties of concrete. Electrical properties are useful for determining the dissipated or required currents when treatment of concrete is applied.
- Choose a chemical solution to be used in protecting concrete so that reactions do not occur which cause further damage.
- Cost factor. Cost is the most important factor in carrying out maintenance where comparisons can be made between one treatment system and another by considering the risks and benefits.

To keep concrete in stable strength, continuous maintenance is required. Concrete teratment should be carried out while concrete is being made or concrete that has been built for a long time. The following are concrete curing methods that have been applied in an effort to extend the life of concrete building structures. Table 10.1 list principles and methods to protect concrete structure from deterioration according to EN 1504–9; Lech [1].

Table 10.1 Principles and methods for the repair and protection of concrete according to EN 1504–9; Lech [1]

Work principle	Method	Polymer materials
Resist seepage	Hydrophobic Injection Coatings	Silanes, siloxanes, Epoxy-acrylic-polyurethane, resins
Keep moisture	Electro-chemical treatment Coatings	Epoxy, polyeutherene, acrylic-resin
Restore concrete condition	Casting Replace elements Spray new concrete	Polymer mortars. Polymer cement mortar
Increases concrete strength	Add steel reinforcement Binding with plates Reinforcing cement injection	Polymer mortars. Polymer cement mortar
Physical endurance	Adding mortar Coatings	Polymer concrete, polymer cement-concrete
Chemical resistance	Coatings Add mortar	Polymer mortars. Polymer cement mortar

10.2 Re-alkalization of Concrete

Corrosion of steel reinforcement has become the main cause in reducing strength of concrete used as building structures. Therefore, corrosion prevention must be the main concern in extending the life of the building and preventing sudden damage. Factors causing corrosion of steel reinforcement are the entry of water and chemical elements (CO_2, H_2S, S_2, Cl) into the concrete which can damage the protective alkaline layer of reinforcing steel. At that time the concrete was no longer passive (alkaline) and turned into sour condition [2].

Steel surfaces that are exposed to an acidic environment will experience ionization (corrosion) continuously eventually the concrete will lose its strength. One way to prevent corrosion of reinforcement is to restore the concrete to be passive (re-alkalinated). Re-alkalination is an electrochemical method that uses an electric current to force negative ions going out from the concrete. It can be conducted by inecting a negative charge to the reinforcing steel and a positive charge outside the concrete (see Fig. 10.1). In this way the negative ions will go to the positive pole towards the concrete surface and make the steel back to alkaline [2].

The purpose of re-alkalization is to restore the alkalinity of concrete which has become acidic due to the inclusion of CO_2 or Cl^- ions. Re-alkalination reshapes the passive environment around the reinforcement by creating an electric field around the concrete.

By monitoring the voltage, concentration, type of electrolyte, polarization time, current, and corrosion rate of the steel, the effect of re-alkalization can be determined.

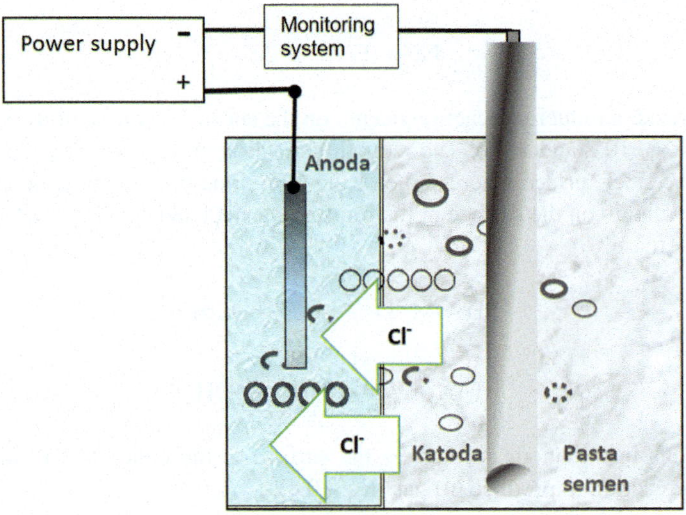

Fig. 10.1 Schematic of the re-alkalination process where a negative current is applied to the reinforced concrete electrochemically, the workpiece is changed in potential (polarization) and the current suppying from the power supply is recorded

In order to control the realkalization process and achieve the best effect, several studies on the process parameters of the re-alkalization effect have been reported. Matsumoto et al. [3] used Li_3BO_3 and Na_2CO_3 solutions as electrolytes in the re-alkalization process. The results showed that Na_2CO_3 solution as an electrolyte causes expansion of concrete (alkaline aggregate reaction), but the addition of Li_3BO_3 solution as an electrolyte can reduce the alkaline aggregate reaction. K_2CO_3 solution is known to be effective in preventing the alkaline silicon reaction. Bastidas et al. [4] studied the use of re-alkalization of carbonated concrete. In his research shows that re-alkalization can reduce the impact of steel carbonization in concrete. And several other researchers have proven that re-alkalization is a method that can be used to prevent corrosion of steel in concrete [5–7].

Basically, it can be concluded that the re-alkalization process can make concrete return to alkaline conditions. However, it must be noted that there are other impacts that can be detrimental to concrete, namely the alkaline-aggregate reaction, which can worsen the condition of the concrete. Thus, in carrying out the re-alkalization process, important parameters that must be considered are the type and concentration of electrolyte, electrification time, and the physical state of the surface of the steel bar to which the re-alkalization process will be applied.

10.2.1 Principle of Re-alkalization

The re-alkalization process causes the corrosion reaction on the reinforcing steel (Eq. 10.1) not to occur [8].

$$Fe \rightarrow Fe2^{+} + 2e^{-} \tag{10.1}$$

As Instead, a reduction reaction occurs on the surface of the reinforcing steel, as follows (Eqs. 10.2 and 10.3). The reduction reaction at the cathode produces OH^- ions on the steel surface where some of the OH^- ions flow to the concrete surface and some remain on the surface of the reinforcing steel and cause the pH around the steel to increase.

$$2\,H_2O + O_2 + 4e^{-} \rightarrow 4OH^{-} \tag{10.2}$$

$$H_2O + 2e^{-} \rightarrow H^{+} + OH^{-} \tag{10.3}$$

OH- ions in the anode located on the surface of the concrete will undergo an oxidation reaction to produce O_2 gas (Eq. 10.4).

$$4OH^{-} \rightarrow 2\,H_2O + O_2 + 4e^{-} \tag{10.4}$$

$$2\,H_2O \rightarrow O_2 + 4\,H^{+} + 4e^{-} \tag{10.5}$$

As seen from above, the re-alkalization effect of concrete is mainly based on two aspects: the electrochemical reaction of the cathode and the electro-osmosis of the alkaline solution. Oxidation reactions also occur at the anode against water on the surface of the concrete. This system can be used to overcome the problem of carbonation without the need to replace contaminated concrete so that cost savings are obtained.

Another way treating the concrete with the re-alkalinization method and avoiding the silica-alkali reaction is by covering the concrete with an alkaline solution. Reinforced concrete is wrapped with a wet cloth containing an alkaline hydroxide solution inserted into the pores of the concrete. Alkali concentration, temperature, water and time must be measured to prevent excess alkaline ions from entering the concrete pores which can cause swelling. In this way, the silica gel contained in the concrete can be removed [9].

10.3 Dechlorination (Chloride Removal)

Chloride in concrete can be removed by electrochemical method. It is conduced by applying the anode and electrolyte on the surface of the concrete structure which is connected to a DC voltage regulator. This method makes the reinforcing steel a cathode. As a result of connecting the negative current to the steel reinforcement in the concrete, anions (negatively charged ions) migrate towards the anode (concrete surface), including chloride ions which also migrate away from the reinforcing steel. The speed of reduction of chloride ions depends on the amount of current used. In general, the current used to remove chlorides is 50 to 500 times higher than that used for cathodic protection [2]. It should be noted that the movement of ions through the concrete has the potential for damage to the steel reinforcement. The chemical reaction that occurs when transferring a negative current to the steel is a reduction reaction in which hydrogen gas and OH^- ions will be produced (Eqs. 10.6 and 10.7). the more negative the current is used the more H_2 gas and OH^- ions are produced [2].

$$O_2 + 2\,H_2O + 4e^- \rightarrow 4(OH)^- \tag{10.6}$$

$$2\,H_2O + 2e^- \rightarrow H_2 + 2(OH)^- \tag{10.7}$$

The reaction in Eq. 10.6 is very slow because it depends on the availability of O_2 in the concrete where the O_2 content is very limited. So the most production is H_2 gas. It is feared that the results of this reaction can change the cement to become soft and reduce the steel–concrete bonding force. Excessive H_2 gas has the potential to cause hydrogen embrittlement and cause tensile stress in concrete. The dechlorination process produces OH- ions (Eq. 10.6) which can create an alkaline atmosphere on the surface of the steel and make the steel become passive again and helps prevent the return of corrosion. The electrochemical reactions that occur are as follows:

$$\begin{aligned} &pH < 4 \\ &2\,H_2O \rightarrow O_2 + 4\,H^+ + 4e \end{aligned} \tag{10.8}$$

$$\begin{aligned} &pH > 72 \\ &Cl \rightarrow Cl_2, +2e \end{aligned} \tag{10.9}$$

Equations 10.8 and 10.9 show that Cl_2 gas and O_2 gas will form on the surface of the concrete if a positive current is connected to the electrode at the surface of the concrete. Excessive chlorine gas content can cause pollution and O_2 gas is known to cause damage to concrete. Thus, it is necessary to carry out proper monitoring when using current for the purpose of dechlorination. In order for the dechlorination process to run effectively, several things that need to be considered are as follows [2].

- *Not for prestressed steel*

The process of removing chloride will produce hydrogen in the reinforcing steel. Prestressed steel is weak against H_2 gas which can cause hydrogen embrittlement in the prestressed steel structure.

- *Alkali-reactive aggregate*

The chloride removal process can trigger an alkali-reactive aggregate reaction. Excessive current will produce excessive OH^- ions which can weaken the steel–concrete bond.

- *No serious damage to the concrete*

Chloride removal will work well if there has not been severe damage to the concrete. If a lot of concrete is peeled off, it causes the concrete resistance to become unstable which results in uneven current distribution. If chloride removal is to be carried out on a fractured structure, it is advisable to apply electrolyte between the cracked or damaged areas.

- *Concrete not covered with a non-conductive layer*

The presence of a surface coating or non-conductive coating will result in a weak current due to the high electrical resistance.

- *Electrical Continuity*

The chloride removal process requires a good electrical capacity in order to obtain an even current distribution throughout the structure. Current will not flow steadily in heavily corroded steel.

- *Low concrete electrical resistance*

If the electrical resistance of the concrete is too high, the distributed current becomes very low and a long time is required for chloride removal. Concrete resistance can be determined in the field using a resistance meter [2].

10.4 Curing

Curing is the process of maintaining temperature and humidity conditions at the desired condition so that the concrete avoids cracking due to the hydration process. When concrete undergoes a hydration reaction, heat release occurs quickly, allowing the concrete to crack if not cured. When concrete undergoes the drying process, some water will be lost due to reaction with cement which leaves hollow spaces in the concrete. Proper curing can increase compressive strength, avoid cracking and improve concrete porosity [10]. Curing can be done in several ways, namely:

- *Curing using water*

The curing method with water is the most common method because it is easy to do. This method can withstand the presence of moisture in the concrete during the hardening period. Curing with water media is done in various ways, namely by placing the concrete into a pool of water during the hardening process, irrigating the concrete construction with water, and spraying water on the surface of the concrete castings.

- *Curing using a membrane coating*

This method aims to prevent the loss of water from the concrete surface. This method is carried out if there is not enough water at the casting location. Therefore, maintenance of the membrane can be done in the following way: placing a wet cloth over the surface of the concrete so that the concrete is always damp. Installing a plastic cover over the concrete surface. This is so that the water does not evaporate much. Using chemical compounds. Concrete is coated with a chemical compound that forms a thin liquid membrane on the surface of the concrete so as to prevent evaporation.

- *Use of heat (steam curing)*

In steam curing, concrete is placed in a chamber or room filled with steam (water vapor). In steam curing, the strength of concrete can be increased because the steam will provide additional heat and moisture to the concrete. When concrete is heated to a higher temperature, the hydration process takes place rapidly. Concrete subjected to this hardening technique tends to dry more uniformly and much more rapidly than other processes. Steam curing will be more effective if conducted at high temperature and pressure.

- *Curing concrete with CO_2*

Carbonation curing causes a chemical reaction to occur between the cement binder and carbon dioxide at the initial setting. Reactions occur between cement in the anhydrous phase (C_3S or C_2S) and CO_2 during the concrete compaction process. Curing by CO_2 can also occur between hydration products such as calcium hydroxide ($Ca(OH)_2$) and calcium silicate hydrate ($3CaO{\cdot}2SiO_2{\cdot}3H_2O$, or CSH) and carbon dioxide in concrete. This carbonation treatment was able to improve the strength of concrete and showed an increase in resistance to sulfate attack [11].

10.5 Coatings

Damage to concrete begins with the surface quality of the concrete. Thus the treatment of concrete is the key to determining the service life of concrete. The durability of concrete is determined by porosity, dust and dirt pollutants that stick to the surface of the concrete. In order for the concrete to have a long life, the effort that needs to be done is to keep the surface of the concrete clean from chemicals and free from defects. The advantages obtained if the concrete surface conditions are treated are: increased surface strength, dust free, increased abrasion and chemical resistance, ease of cleaning.

The humidity of the concrete needs to be maintained so that the amount of moisture does not accumulate on the surface of the concrete. Steam that sticks to the surface of the concrete can dissolve pollutants which then penetrate into the reinforcement through the pores of the concrete. Mold growth and salt buildup are triggered by the moisture in the concrete. Concrete that is exposed to the outside air also experiences repeated hot and cold cycles so that the aggregate can become weathered and cracked. To prevent damage as mentioned above, the surface of the concrete wall must be treated with paint as a barrier layer to prevent water penetration into the concrete [12].

10.5.1 Coating Function

Barrier. A layer that forms a barrier between concrete surfaces and prevents entry of harmful substances into the concrete body through transport mechanisms such as absorption, capillary action and diffusion. Basically, a coating is a film-forming substance in the curing process. Barrier example: acrylics, epoxies, coal tar epoxies.

Penetrant inhibitors. Is a primer that can dissolve in water or solvents which functions to reduce the speed of corrosion by forming a thin layer that blocks electro-chemical reactions. Example: Silanes—siloxanes.

Galvanic coating. Made of zinc which can provide galvanic or cathodic protection to ferrous metals. Zinc is more active than iron so that if it is connected to one another, zinc will undergo an oxidation reaction and iron will undergo a cathodic reaction.

10.5.2 Coating Materials

One of the method to improve concrete durability is by applying polymer coat in the surface of concrete structure. Polymer coat has a function to create a barrier for the harmful substances to transform into the concrete structure. Table 10.2 is several types of polymers used for concrete coatings [13].

Table 10.2 Types of polymers used for concrete coatings

Polymer	Chemical structures	Polymer characteristics
Polyurethanes	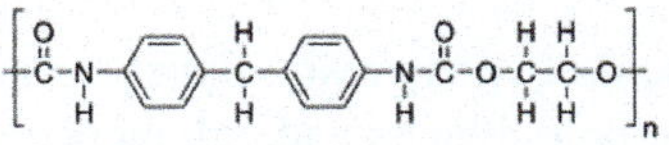	Polyurethanes are used in a variety of applications, including: Building insulation. Refrigerators and freezers. Furniture and bedding. Footwear. Automotive. Coatings and adhesives
Polycarbonate	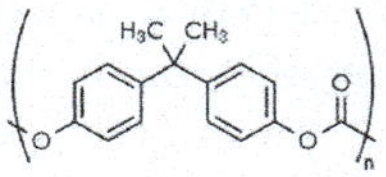	Hard coatings are applied to polycarbonate eyeglass lenses and polycarbonate exterior automotive components
Pp (polypropylene)	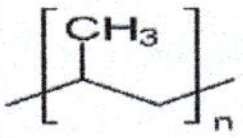	Polypropylene is similar to polyethylene in many aspects, especially in solution behavior and electrical properties. The methyl group improves the mechanical properties and thermal resistance, although the chemical resistance decreases. The properties of polypropylene depend on the molecular weight and distribution of molecular weights, crystallinity, types and proportions of comonomers (if used) and isotacticity
Alkyds	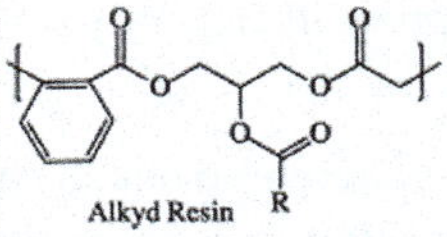	Alkyd is used in paints, varnishes and molds for foundries. They are the predominant resins or binders in most commercial oil-based coatings
Acrylic ($C_5O_2H_8$)n	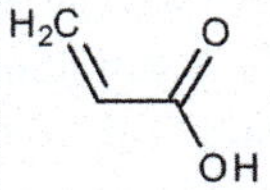	Acrylic is a transparent plastic material with outstanding strength, rigidity and optical clarity. Acrylic sheets are easy to manufacture, bond well with adhesives and solvents, and are easy to shape with heat. It has superior weathering properties compared to many other transparent plastics

Acrylics. Acrylic paints are fast-drying paints made from pigments suspended in an acrylic polymer emulsion**.** Acrylic paint can be dissolved by water, and become waterproof after drying.

Alkyds. Alkyds are natural oils (soybean, tungsten, styrene) that have been chemically modified to improve drying properties and increase chemical resistance. Alkyd-phenolics are used as primers, and alkyd-silicones are used as topcoats for structures exposed to atmospheric conditions. Alyds is not suitable for use on alkaline surfaces or environments (concrete or stone).

Bituminous. Bituminous coatings are heavy petroleum products applied as coatings. Bituminous has moisture and chemical resistance properties but is not resistant to solvents.

Epoxy. Epoxy is the basic component of epoxy resin. Epoxy resins, also known as polyepoxides, are a class of reactive prepolymers and polymers containing epoxide groups. Epoxy is widely used as an adhesive or coating material.

Inorganic zinc primer. This coating is exclusively a primer as it provides galvanic or cathodic protection to the steel substrate. Inorganic zinc is used for structures exposed to the atmosphere or submerged.

Polyurethane. Polyurethane is a subset of urethane. Polyurethane is used for the top layer of coating to protect the structure from direct sunlight or UV. Polyurethane is suitable for partially submerged structures or fluctuating weather conditions.

Urethane. Urethane keeps concrete moisture exposed to the atmosphere and can be used on damp concrete surfaces. Urethane is suitable for coating structures exposed to the open air.

10.6 Additives

Based on its function, additives that can improve the quality of concrete can be classified into five main categories as follows:

- *Mixed inhibitors.* This substance is added to concrete so that the hydration process occurs slowly. This substance is useful if casting is done in hot air. In the presence of this substance, the speed of hydration will be reduced and the possibility of cracks can be avoided. Hydration inhibitors can simultaneously be useful in reducing the amount of water needed when setting with reference to ASTM C494/C494M-17. This specification provides guidance on the materials and test methods that need to be used in preparing chemical admixtures to be added to cement concrete mixes. There are seven types of mixtures (AS) that can be used for seven types of intended use [14].
- *Acceleration of the mixing process.* This substance is able to shorten the concrete casting time, can be used in cold weather, and finishes the surface earlier. The use of this substance has a negative impact in the form of large concrete shrinkage. An example of a drying accelerating agent is NaCl salt. The use and method of NaCl application are regulated by ASTM D 98 standard. The amount of NaCl used must be in the recommended amount, if using excessive NaCl in the concrete mix can cause stiffness, increase drying shrinkage and cause corrosion of steel reinforcement.

- *Superplasticizers*. Is a substance that can reduce the water content in the manufacture of concrete. The dosage required varies depending on the type of concrete mix to be made. Superplasticizers can produce high strength concrete and improve flowability.
- Water-reducing admixtures require less water to make slump concretethe same, or increase the slump of concrete at the same moisture content.
- *Air-entraining agent*. Is a chemical substance added to concrete to form air bubbles in the concrete. With the presence of air bubbles in the concrete, it can provide space if there is an expansion of the volume of water in the concrete to avoid cracks.
- *Adhesive*. Adhesives are added to increase the strength of the concrete and reduce the pores of the concrete. The types of adhesives commonly used are: polyvinyl chlorides and acetates, acrylics and butadiene-styrene co-polymers.
- *Water resistant substance*. Water-resistant compounds added to concrete for example: soaps, butyl stearate, mineral oil and asphalt emulsions. This material is used to reduce the amount of water penetration into the larger concrete pores.

10.7 Rust Cleaning

Steel if it is in the open will experience corrosion. In open areas steel is exposed to a wet environment where the air contains aggressive elements in the form of chloride or CO_2. These elements have the potential to condense on the steel surface and cause the corrosion process to take place. Rust in steel must be cleaned before casting so that the adhesion of steel and concrete increases. When steel undergoes an initial process of corrosion on its surface, the corrosion will continue even though casting has been carried out. Corrosion will take place very quickly if the concrete becomes acidic or water and oxygen enter the concrete.

10.7.1 How to Clean Rust

Things that need to be considered in determining the rust cleaning method are: severity level, corrosion products, accessibility and cost of the cleaning process. Methods for cleaning rust include the following:

- Wire brushing. Brushing of the reinforcing wire can be done manually or by machine. Brushing with wire manually is a rather difficult and time-consuming process, so it is not recommended for steels that experience heavy corrosion levels.
- Use of high-pressure water jet sprayers. The use of a stream of water can be used to clean rust evenly. Iron sand is often used as an abrasive medium so that the cleaning process takes place quickly. There are several types of abrasive media

used to clean rust, for example by using a metal shot, with sand, glass balls, plastic, dry ice and balls from plants.

- Using chemicals. Strong acids (HCl) and strong bases can be used to clean rust on reinforcing steel. They work by dissolving rust on the surface of the steel. However, this caustic chemical is also corrosive to most other substances, and is very dangerous to use. Weak acids (oxalic acid or EDTA) can also be used to clean rust. Weak acids react with rust less strongly than strong acids, although they are safer to use or easier to dispose of.
- *Dry ice blasting* i.e. using dry ice (solid CO_2). The use of dry ice is beneficial because it does not harm the environment, is relatively clean because it does not pollute the air and does not damage the steel being cleaned. Dry ice uses CO_2 which is cooled to a temperature of around −70 °C so that the CO_2 gas turns into a solid. With the help of a compressor, this dry ice is sprayed onto the corroded concrete surface. When dry ice comes into contact with steel at high speed the dry ice will melt and have a larger volume so that the rust on the steel will dissolve.
- Laser cleaning. The use of lasers has the advantage of precision and low levels of air pollution. Lasers are classified as green technologies that do not use chemicals. Dust generated during cleaning can be sucked up immediately. The use of lasers also reduces labor costs and no routine maintenance is required.

References

1. Czarnecki, L.: Polymer-concrete composites for the repair of concrete structures. MATEC Web Conf. **199**, 01006 (2018)
2. Bennett, J., Schue, T.J.: Chloride Removal Implementation Guide, Strategic Highway Research Program National Research Council. ELTECH Research Corporation Fairport Harbor, Ohio (1993)
3. Matsumoto, K.I., Ueda, T., Ashida, M., Miyagawa, T.: Study on re-alkalization with electrolytes containing lithium ion. Int. J. Mod. Phys. B **17**, 1446–1451 (2003)
4. Bastidas, D.M., Cobo, O., González, J.A.: Electrochemical rehabilitation methods for reinforced concrete structures: advantages and pitfalls. Corros. Eng. Sci. Technol. **43**(3), 248–255 (2008)
5. Miranda, J.M., González, A.C., Otero, E.: Several questions about electrochemical rehabilitation methods for reinforced concrete structures. Corros. Sci. **48**(8), 2172–2188 (2006)
6. Redaelli, E., Bertolini, L.: Electrochemical repair techniques in carbonated concrete. Part II: cathodic protection. J. Appl. Electrochem. **41**(7), 829–837 (2011)
7. Zhang, J., Yan, L.C.: Electrochemical realkalization and combined corrosion inhibition of deeply carbonated historic reinforced concrete. Corros. Eng. Sci. Technol. **48**(1), 28–35 (2013)
8. Zou, Z., Yang, G., Chen, F., Long, F., Li, Q., Jiang, R., Zhang, H.: Re-Alkalization effect experiment and a new re-alkalization control. model of carbonated concrete. Adv. Mater. Sci. Eng. (2022) Hindawi, 8 Nov.
9. Kawabata, Y., Yamada, K., Sagawa, Y., Ogawa, S.: Alkali-wrapped concrete prism test (AW-CPT)–new testing protocol toward a performance test against alkali-silica reaction. J. Adv. Concr. Technol. **16**, 441–460 (2018)
10. Du, X., Li, Z., Tong, T., Li, B., Liu, H.: Isothermal drying process and its effect on compressive strength of concrete in multiscale. Appl. Sci. MDPI (2019)

11. Shao, Y.: Beneficial Use of Carbon Dioxide in Precast Concrete Production, Final Report, March, National Energy Technology Laboratory (2010)
12. Teknos, O.: Handbook for the Surface Treatment of Concrete (2014). www.tecnos.com
13. Ramesh, M.N.: Talrak Construction Chemicals Pvt. Ltd., Bangalore (2017). https://www.nbmcw.com/product-technology/construction-chemicals-waterproofing/waterproofing-repair-chemicals/concrete-protection-coatings-for-reinforced-concrete-structures.html
14. Gopal, M.: (2020). https://theconstructor.org/concrete/concrete-chemicals-admixture-types-uses/5978/
15. ASTM C-494/C-494M-17.: Standard specification for chemical admixtures for concrete.
16. ASTM D-98.: Standard Specification for Calcium Chloride, June 1 (2015)
17. Fontana, M.G.: Corrosion Engineering, 3rd edn. McGraw-Hill, New York (1986)

Chapter 11
Cathodic Protection of Steel Reinforcement

Abstract This chapter discusses cathodic protection as a highly effective approach to combat corrosion of steel reinforcement in concrete structures. It thoroughly examines the considerations involved in designing cathodic protection systems specifically tailored for reinforced concrete applications. The role and significance of sacrificial anodes and impressed cathodic protection in protecting the steel reinforcement are explored in depth. The concept of cathodic current requirement protection, which plays a crucial role in determining the optimal amount of current necessary for effective corrosion prevention, is extensively discussed. Moreover, the chapter investigates the occurrence of peeling off or cathodic disbonding, which can occur under specific conditions during cathodic protection. The monitoring of cathodic protection systems and the potential challenges associated with stray current and current disturbance are also carefully examined. Additionally, the chapter introduces anodic protection as a viable alternative for corrosion prevention in reinforced concrete. By thoroughly exploring these topics, this chapter offers a comprehensive overview of cathodic protection, providing valuable insights and knowledge for engineers, researchers, and professionals engaged in corrosion prevention within the field of concrete structures.

11.1 Introduction

Cathodic protection is a method to protect building structures so that they last longer. Structures such as buildings, bridges, dams, water embankments, and tunnels that are exposed to air or water for a long time will suffer damage. Concrete material will easily experience a decrease in quality due to chemical interactions between the constituent materials of concrete and the surrounding environment. Free air, especially in industrial areas containing carbon dioxide, sulfide, water vapor, dust, and water can enter through the pores of the concrete to form compounds with the chemical elements of concrete (calcium, aluminum, silicon, and magnesium). The chemical elements making up concrete will become active if they interact with pollutant elements, which results in the weakening of the concrete bonding force and potentially causing cavities, crack growth, and weathering. The damage will

Y. P. Asmara, *Concrete Reinforcement Degradation and Rehabilitation*,
Engineering Materials, https://doi.org/10.1007/978-981-99-5933-4_11

be exacerbated if the steel reinforcement in the concrete corrodes. The corrosion reaction on the steel surface increases the volume of the steel, which produces a compressive force on the concrete. Both corrosion and reduced quality of concrete work reduce the life of reinforced concrete buildings.

Efforts to extend the life of reinforced concrete structures are carried out by inhibiting and preventing corrosion in the reinforcement. There are several methods for preventing corrosion, namely cathodic protection (CP), anodic protection, adding inhibitors, and modifying the properties of concrete [1]. Currently, protection systems with CP are widely used to prevent the corrosion of steel reinforcement. Cathodic protection has demonstrated a success rate of 90% [2]. Besides being useful for preventing corrosion, CP can also be used for the rehabilitation of damaged and old structures. Even though CP costs a lot, it is still relatively lower compared to the risk if the building structure is not protected. With CP, the high cost of repairing and removing damaged concrete can be avoided.

Cathodic protection can be grouped into two basic types, namely impressed current (ICCP) and sacrificial anode (SA). Impressed current cathodic protection is carried out by providing direct current (DC) to the steel reinforcement [3]. The DC from an external source is used to change the reinforcement potential to the cathodic potential. The protection system using SA is carried out by supplying electrons that come from other materials that are more negative, which are usually more negative (50–200 V) than the corrosion potential of the protected object. In this way, the reinforcement is no longer subjected to oxidation. Cathodic protection supplies external energy to the steel surface embedded in the concrete by forcing all reinforcing steel to act as a current-accepting cathode instead of providing electrons. Electrons from outside are distributed throughout the protected metal parts. Thus, the calculation of conductivity and electron density becomes an important parameter in designing the protection of concrete structures. Conductivity measurement of soil, concrete, and water ensures that the current will be uniformly distributed in all protected structures [4].

This chapter presents cathodic protection of steel reinforcement which consist of: Designing cathodic protection for reinforced concrete; Cathodic current requirement protection; Peeling off (cathodic disbonding); Monitoring of cathodic protection; Stray current and current disturbance; and Anodic protection [5].

11.2 Designing Cathodic Protection for Reinforced Concrete

Designing the protection of reinforced concrete structures requires data on the characteristics of the metal being protected and the location where the structure is located. Each location has different characteristics in terms of electrical properties, water content, the presence of similar installations, and weather changes. These factors will determine the type of protection that must be selected. The following criteria are used to determine the type of reinforced concrete protection:

- Protection plan time.
- Electrical continuity. Electrical circuits must always be closed so that the system can function continuously.
- Calculation of chloride concentrations and chemical elements around the site. If the levels of chloride and chemical element concentrations are high enough, CP is the only method that is feasible to use.
- Possible alkali-silica reaction. Cathodic protection increases the alkalinity on the steel–concrete surface, thereby potentially accelerating the alkali-silica reaction.
- Cost and ease to install, monitor, and repair.

11.2.1 Impressed Current Cathodic Protection

In this method, a metal structure is used as the cathode by connecting the negative terminal of the current source to the structure and the positive terminal to the inert anode. Figure 11.1 shows principles of steel reinforcement protection with the ICCP system. The anode is another metal that is placed near the structure and is electrochemically connected to the structure. The tools needed to design the ICCP are:

- Rectifier (source and DC provider).
- Anode. References are the basis for determining the amount of potential protection.
- Electrolytic media (easily conducts electric charge).
- Protected metal (steel reinforcement).
- Electrical circuit that stabilizes the magnitude and direction of the current.
- Evaluation and control devices during protection (probes, reference cells, and settings).

They are can be described in details as follow.

- A rectifier is a device that converts alternating current (AC) into one-way DC. The device can also be used to automatically adjust the amount of current needed to protect steel. The rectifier keeps the potential difference between the protected structure and the anode constant.
- Anode. The ICCP uses an anode which is useful for making the structure always cathodic relative to the anode installed. The materials used as anodes vary, and an example is a material made of inert mixed metal oxide (MMO). Mixed metal oxides are mixed materials made of various types of metals and ceramics. The main alloy used is titanium (titanium dioxide (TiO_2) *and* tantalum oxide (TaO_5) mixed with ceramic materials like ruthenium oxide (RuO_2) or iridium oxide (IrO_2). Mixed metal oxide has high efficiency but the price is high. Scrap steel anode is preferred due to its low cost although it has low efficiency. Titanium,

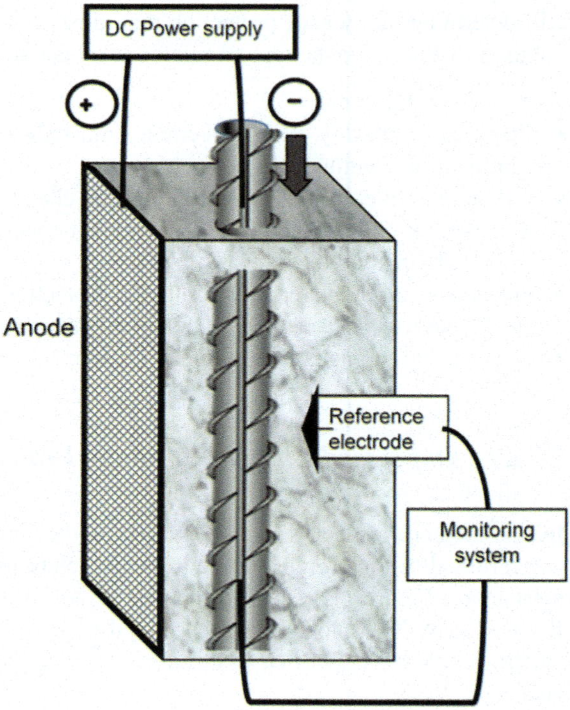

Fig. 11.1 Scheme of steel reinforcement protection with the ICCP system

graphite, and silicon-iron-coated MMOs are the most widely used for underground applications. Graphite and silicon-iron are brittle and must be handled with care. The anode is usually surrounded by a coke-like deposit to increase the electrical contact between the anode and the surrounding soil. For use in marine locations, lead-silver alloys, platinized titanium, or platinized niobium are commonly used. Table 11.1 presents the anode materials used in ICCP. During ICCP protection, the reaction that occurs at the anode is the formation of chloride gas if the soil has high salinity and the formation of oxygen if the salinity is low. If the shelter contains water and carbon dioxide, carbonic acid formation will occur. This reaction at the anode will create high acidity on the surface of the anode; therefore, an anode material that is resistant to acid attack is needed.

- Groundbeds. Implementation of ICCP in the ground requires a medium that can fully utilize the current coming from the rectifier so that all surfaces of the structure receive a large current. The media are called groundbeds, which are containers containing graphite or bentonite. Groundbeds can be classified into several types, namely vertical, horizontal, or deep wheel. Groundbeds help reduce soil resistance, extending the life of the anode by allowing the anodic reaction to occur on the surface only. Groundbeds are made porous so that the gas produced by the corrosion reaction can escape.

Table 11.1 Anode materials used in the ICCP system [2]

Material	Consumption rate (kg/amp-yr)	Typical current density (amp/m^2)
MMO-coated titanium	10^{-6}	100–600
Platinized titanium or niobium	10^{-5}	100–1,000
Silicon-chromium-iron	0.4	10–40
Magnetite	0.1	3–60
Steel	7–9	0.1–1
Lead-silver alloy	0.05–0.1	150–200
Graphite in coke	0.2–0.5	10–40

- Cable and connectivity. Connection cable management is very important and determines the success of ICCP. The factors that must be considered in designing the ICCP are that the cable must be well insulated, ideally without joints and free from exposed surfaces, and always protected from damage during installation. Cables are usually brazed or welded to the structure using thermite welding to ensure a strong bond. To ensure that the connection is always watertight, it is necessary to provide an insulating patch or a coating.

The advantages and benefits of ICCP protection are:

- Easy to achieve the required current (unlimited availability).
- More flexible voltage and current settings.
- Suitable for use in locations with high resistivity, large objects, and uncoated materials.

The weaknesses of ICCP protection are:

- Expensive installation costs for small systems.
- Higher maintenance costs for small systems.
- May cause adverse effects (disturbance) to nearby structures.

11.2.2 Sacrificial Anode

In this method, the building structure is made into a cathode by connecting it with a material that is more actively oxidizing or more susceptible to corrosion (see Fig. 11.2). In other words, SA protection system is associated with a material that is more electronegative than the metal structure to be protected. Table 11.2 lists Sacrificial anode (SA) materials commonly used to protect steel.

The selection of anodes is based on economic and engineering considerations, such as the amount of electrical energy that can be used. The anode current capacity indicates the number of ampere-hours that can be supplied by each kilogram of the

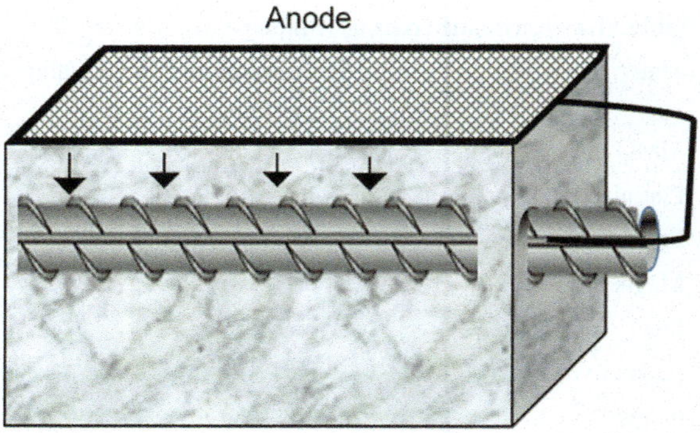

Fig. 11.2 Scheme of protection of steel reinforcement with a sacrificial anode system (i.e., CP sacrificial anode)

Table 11.2 Materials commonly used as sacrificial anodes [2]

Anode type	Potential (volts) versus $Cu/CuSO_4$	Practical consumption rate (kg/amp-yr)	Practical capacity (Ah/kg)	Application
Zinc	–1.1	11	800	Marine and low-to-medium resistivity soil or water
Magnesium	–1.55 (Std) –1.80 (High pot)	7 4	1,100 1,200	Buried and non-saline water
Aluminum	–1.1 –1.1 to –1.17	3.2 (Al–Zn–Hg) 3–4 (Al–Zn–In)	2,800 1,700–2,600	Marine only

anode. It should be noted that as the anodes in SA protection are also self-corroding, the efficiency is less than 100%.

The characteristics of the anode are expressed by the speed consumed per kilogram per ampere-year. This data provides information on the weight of the anode that will be consumed for every one ampere used for one year. This figure, calculated by dividing the number of hours in one year (8.760) by the capacity in ampere-hours (Ah) per kilogram, is useful for calculating the expected life of the installation. Table 11.2 shows three types of SA materials and their properties. It should be noted that there will always be anode material left when used. The residual anode factor is around 0.85. Hence, the actual number of anodes required for protection is the theoretical number divided by 0.85.

The advantages of the SA method are:

- No external power source is required; hence, it can be used remotely.
- Lower installation costs.
- Minimal maintenance is required.
- Rarely causes adverse effect (interference) on other structures.

The limitations of the SA method are:

- Movement and the amount of current available are limited.
- Not effective in environments with high resistivity.
- Not effective for use on structures with poor coating conditions.
- The anode life tends to be relatively short, depending on the area being protected.
- Its existence is easily overlooked by maintenance personnel.

Table 11.3 List of electromotive force series indicating the reactivity of a metal. The more negative the potential value of the metal, the more active the metal (easily undergoes oxidation/corrosion) [6, 7, 10].

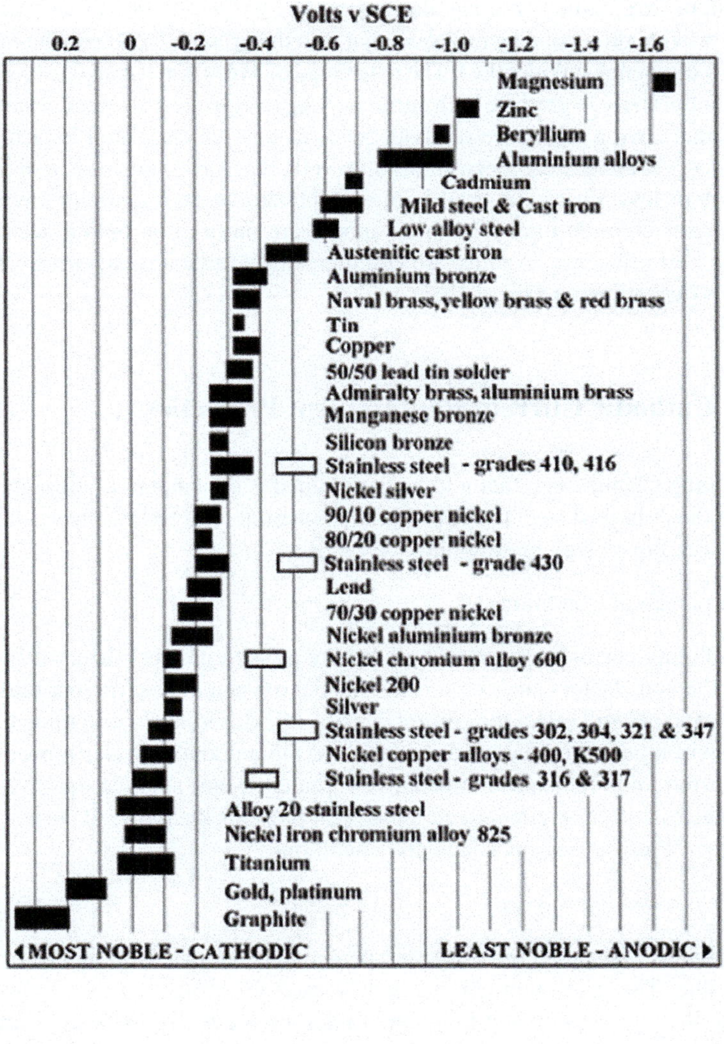

a. Selection of sacrificial anode material

Materials that can be used as SA are materials that have higher reactivity than the protected material (lower reduction potential). Table 11.3 shows the potential values of one metal relative to the potential values of other metals. The potential was measured using a standard calomel reference electrode. The saturated calomel electrode (SCE) is a reference electrode that uses the reaction between mercury and mercury (I) chloride. The potential shown is the reduction potential. The more negative the reduction potential value of a material, the easier it is for the material to undergo oxidation reactions. For example, to protect carbon steel in concrete, an anode material that has the potential to be more reactive to steel is needed, namely magnesium (Mg), zinc (Zn), and aluminum (Al). Currently, Mg anodes are widely used to provide protection against steel that is in the ground. However, the Mg anode has a low efficiency (about 50%). The advantage of Mg is that it can give a very negative potential so it can produce high currents. Magnesium can provide steel protection in soils that have a resistance of about 50 Ω-meters which cannot be achieved by Zn anodes. Zn anodes can only be used on soils that have a resistance of about 10 Ω-meters or less. Zn can be used in most applications. In a seawater environment that contains chloride ions, aluminum anodes are more of an option because they have the best efficiency. Anodes are commercially available in a variety of shapes and sizes depending on use.

11.3 Cathodic Current Requirement Protection

In designing CP, the main factor to be determined is the amount of current required to bring the polarized steel to the protective potential. The following are the factors that must be considered in designing CP [2, 6].

- Environmental Conditions

Steel in highly corrosive seawater will require a much greater current than steel in soil. In the soil, factors such as soil resistance, oxygen concentration, presence of bacteria, and pH will affect the current density. It is different for designing CP in the water environment. In aqueous environments, the amount of water movement can affect current requirements significantly. Turbulent water tends to dissolve oxygen and produces a depolarizing effect. Table 11.4 presents the required current density values for CP in various environmental conditions.

- There is a coating (layer)

The current used to protect the structure being painted (coated) should be of a much lower value. The protection current is only used for protecting the surface of the structure that is not protected by paint. Tables 11.4 and 11.5 provide estimates of the current requirements used to protect various types of conditions that the steel

Table 11.4 Current required for treatment of steel reinforcement structures (*ACA*)

Protection level	Objective	Large current density (mA/m^2)
Corrosion prevention	Prevents corrosion	0.25–2
Controlling corrosion	Reduces corrosion speed	1–7
Cathodic protection	Stops corrosion	2–20

Table 11.5 Current required for cathodic metal protection in various environments (ACA, 2013)

Condition	Required current (mA/m^2)
Effect of environment	
Hot sulfuric acid	400,000
Bare steel in moving seawater	100–150
Bare steel in quiet seawater	50–80
Bare steel in fresh water	40–60
Bare steel in earth	10–30
Reinforcing steel in concrete	5–25
Steel in well-cured, chloride-free concrete	~0.07
Effect of coating	
Poorly coated steel in earth or water	~1
Well-coated steel in earth or water	~0.03
Very well-coated steel in earth or water	<0.01

is coated on. The better the condition of the coating, the less current is required. When the pipe lining is still good, it is possible to cathodically protect sufficient pipe for 80 km or more. Providing continuous current to the surface of the structure can cause the formation of hydroxyl ions, which precipitate sodium chloride (NaCl) and magnesium under the coating surface.

- Other requirements

Other requirements that need to be considered are the presence of current resistance from outside the system, the age of the structure, the installation cost, the degree of difficulty, and the presence of stray currents (disturbances due to other metals in the vicinity of the protection site).

11.4 Peeling Off (Cathodic Disbonding)

Cathodic disbonding is the process of weakening the adhesion power of paint/coating on metal due to CP. The main cause is the reaction product in the form of alkali on the steel surface during cathodic reaction. Cathodic comparison can also be caused

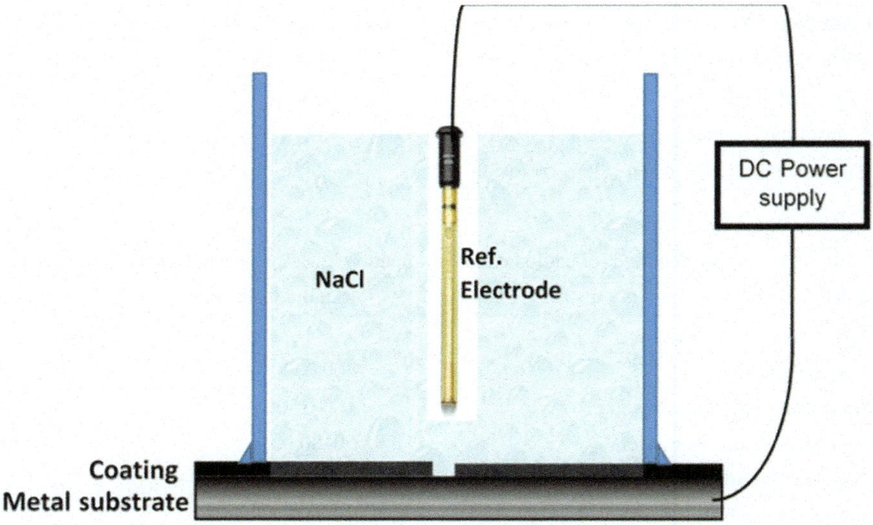

Fig. 11.3 Cathodic disbonding test according to ASTM G8 test standard

by water seepage, the influence of electric fields, and condensation under the coating layer. Another possible cause is the formation of hydrogen when the material is polarized to a more negative potential. The type of coating determines the disbonding rate of the material. Coatings made of epoxy material have shown good resistance, while liquid-based coating materials are often damaged when CP is applied.

Cathodic disbonding resistance is measured using a standard test where holes are made in the coating, and the difference in the size of the holes before and after cathode protection is measured (*ASTM G8 or AS 4352*). Figure 11.3 shows an experimental comparison of coatings due to the effect of CP. Usually, there is a relationship among the width of the disbonding, the time of exposure, and the potential for protection. The more negative the protective potential and the longer the exposure time, the wider the width compared to it. In order to reduce the effect of disbonding, the surface quality needs to be maintained by means of better surface preparation. As the applied potential increases, the rate of cathodic discharge increases; hence, there is an upper limit to the potential for CP, usually around 1.2 V versus a copper sulfate electrode [7].

11.5 Monitoring of Cathodic Protection

Continuous monitoring must be carried out based on established standards to ensure that CP is successful within the planned timeframe. Monitoring can be carried out in two ways, namely checking equipment regularly to ensure that the system is operating properly and periodically to ensure that the CP position is always reached. Equipment checks need to be carried out to ensure that the current always flows in the system and the potential is always within the protection potential. Periodic checks involve measuring the pipe potential at each test point and carrying out further investigations for corrective action if readings indicate inadequate protection.

Pipeline potential measurements are made using a voltmeter that measures the potential difference between the pipe (via the test point lead) and the copper/copper sulfate reference electrode located on top of the pipe. If a pipe is exposed to a stray DC, its potential is constantly changing. To evaluate the protective condition of a pipeline, its potential must be recorded continuously over a specified period of time. There are three types of potential values measured in CP as follows:

- Natural potential

Natural potential which is the potential of the structure before current is applied to the structure. This natural potential is used to determine the value of the corrosion potential. It should be noted that if a structure has been subjected to the application of DC even temporarily, the current balance will be disturbed.

- Active potential

Active potential which is the potential measured when the structure is given CP current. Application of CP will usually change the structure potential in a more negative direction, around −550 to −960 mV.

- Instant-off potential

Instant-off potential which is the potential when the CP potential is turned off instantly. The measurement obtained is actually not only the potential value of the structure that has been polarized but also the potential value due to the CP current flow. This component is known as a decrease in IR. Due to this current flow, the actual measured potential is the sum of the polarized potentials of the structure plus the IR drop. The IR drop can have high values if the coating is damaged, and the values can reach around 500 mV or more. Therefore, to achieve a −850 mV polarized potential on the surface of the structure, the voltmeter value when the CP is activated must be $-850 + (-500) = -1{,}350$ mV.

The potential value when the current is turned off suddenly (instantly) can be used to measure the potential of the actual structure. The value of the instant-off potential is needed to avoid errors caused by voltage changes. When the current is turned off instantly, the measured potential is no longer affected by the IR drop. However, the potential will always change over time. It should be noted that on coated structural surfaces, the time from on to off must be fast (*Fontana*). Figure 11.4 shows values of OFF and ON potential in a CP system. Figure 11.5 shows the results of a potential value survey along a CP protected underground steel pipeline. The readings recorded are natural potential, and on and off potential.

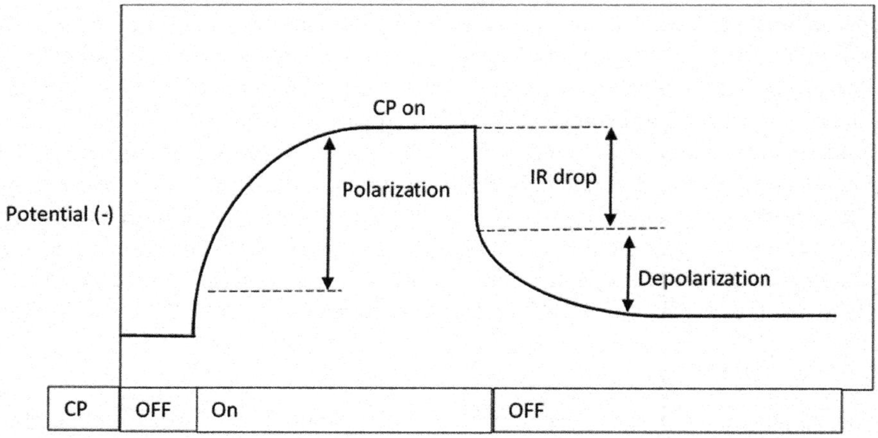

Fig. 11.4 Definition of OFF and ON potential in a CP system [2]

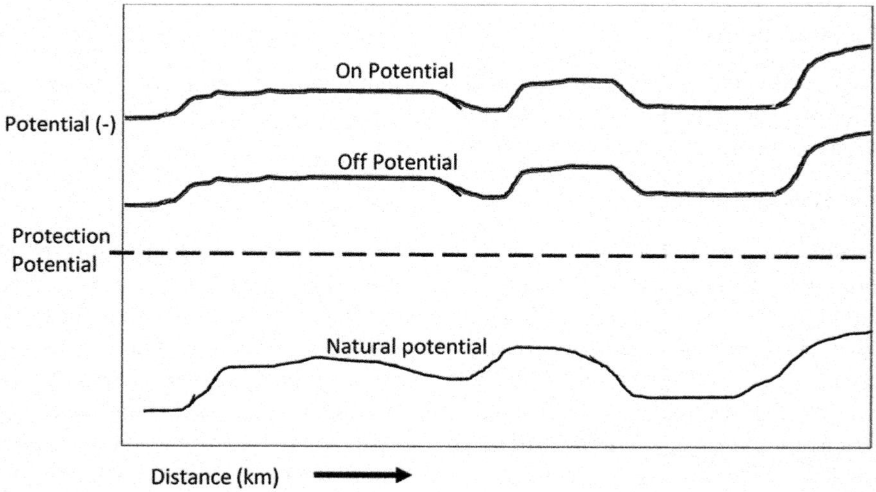

Fig. 11.5 Variation of potential value readings along the path of the structure protected by CP. Potential changes along the path are caused by variations in electrical conditions around the site [2]

11.6 Stray Current and Current Disturbance

Stray current or fault current is an unwanted current that flows in metal structures and is often encountered in CP systems. If this current is large, it will affect the structure to be protected continuously. Examples of stray current sources are currents originating from the electrical system on trains, during welding, other CP systems that are nearby, and magnetic fields. Stray currents can cause corrosion, damage coatings, damage electrical equipment, or cause errors during CP monitoring. Any detected electrical disturbances caused by stray currents are known as interference currents. The most damaging source of stray currents is the existence of railroad tracks and electric trams. Rails that are not isolated from the earth will conduct some of the current to the metal structure that is protected. Current can flow into the structure, causing the structure to polarize when cathodically protected. This condition causes the structure to corrode at a very fast rate.

11.7 Anodic Protection

Anodic protection is a protection method that is carried out by changing the potential of the structure to an anodic position, where at that potential, the material is passive. This protection will only work on metals that produce a passive film when the metal is at an anodic potential, such as iron, chromium, and their alloys. In addition, environmental factors also affect the anodic potential. This protection can only be used in environments where the electrolyte composition supports the formation of a passivation layer. To protect the structure anodically, a potentiostat is used as a current source as well as to maintain the metal at a passive potential. A cathode material made of platinum or silicon iron and stainless steel is required. The size of the electrodes depends on the area covered by the structure being protected. Anodic protection has been proven to reduce the corrosion rate significantly. Anodic protection has been able to reduce the corrosion rate of type 304 stainless steel found in sulfuric acid and NaCl environments by 0.4 μ per year. Anodic protection has the advantage that it can be applied in highly corrosive environments and has low current requirements. Its main limitations are that it is expensive to install and if the protection is damaged, the corrosion rate will be very high. Therefore, it must be used with caution.

Anodic protection has been used to protect storage vessels, process reactors, heat exchangers, and transport vessels containing corrosive solutions, such as sulfuric acid. Most systems are installed in a sulfuric acid vessel. Besides being able to extend its service life, anodic protection can also maintain product quality.

Cathodic protection and anodic protection have similarities and differences, some of which are shown in Table 11.6. Anodic protection can only be used for metals that form a passive film, such as carbon steel and stainless steel, while CP can be used for all types of metals. Anodic protection can be used for highly corrosive environments. Table 11.5 shows that protecting steel in a hot sulfuric acid environment with CP will require 400.000 mA/m^2, whereas if using anodic protection, only about 2% of the current is required. Anodic protection requires more complex equipment than CP, and installation costs are higher. However, running costs can be much lower.

Table 11.6 Comparison between cathodic protection and anodic protection

	Anodic protection	Cathodic protection
Applicable metals	Active–passive	All
Applicable environments	Moderate to aggressive	Mild to moderate
Relative installation costs	High	Low
Relative operating costs	Low	Medium
Important parameters for monitoring	Current	Potential

References

1. Fontana, M.G.: Corrosion Engineering, 3rd edn. McGraw-Hill, New York (1986)
2. ACA 10 – 2013.: Corrosion Technology–Cathodic and Anodic Protection © The Australasian Corrosion Association Inc (2013).www.corrosion.com.au
3. Anderson, G.H.: Proc. Conf. on Cathodic Protection of Reinforced Concrete Bridge Decks, NACE, Houston (1985)
4. Roberge, P.R.: Corrosion Basics-An Introduction, 2nd edn. NACE, Houston, Texas (2005)
5. Joseph, R.D.: Corrosion: Understanding the Basics. ASM International (2000)
6. Shrier's.: Corrosion, 4th Volumes. In: Richardson, T.J.A. (ed.). Elsevier Science (2009)
7. CP1021.: Code of Practice for Cathodic Protection. British Standards Institution, London (1973)
8. ASTM G-84.: Standard Practice for Measurement of Time-of-Wetness on Surfaces Exposed to Wetting Conditions as in Atmospheric Corrosion Testing (2012)
9. AS 4352–2005.: Tests for coating resistance to cathodic disbanding (2005)
10. Jones, D.A.: Principles and Prevention of Corrosion, 2nd edn. Prentice Hall, Upper Saddle River, NJ, USA (1996)
11. Peabody, A.W.: Control of Pipeline Corrosion, 2nd edn. NACE, Houston, Texas (2001)
12. Schweitzer, P.A.: Corrosion Resistance Tables. CRC Press (2004)

The manufacturer's authorised representative in the EU is Springer Nature Customer Service Centre GmbH, Europaplatz 3, 69115 Heidelberg, Germany. If you have any concerns regarding our products, please contact ProductSafety@springernature.com

Printed and bound by CPI Group (UK) Ltd, Croydon, CR0 4YY
01/10/2026
02225877-0012